Beyond the Walls: The Need for Inter-Organizational Knowledge Sharing in South African Construction

Sakina

Table of Contents

1. Introduction

1.1 Introduction

The introduction of this research familiarizes the reader to the topic of the research, its significance to the construction sector and the growing body of knowledge on Knowledge Management (KM) practices, the methodology undertaken to conduct the research and the structure of the document.

1.2 Background to the Study

Research by Kruger and Johnson (2009) showed that organisations within the South African construction sector have largely adopted Knowledge Management as a business strategy to manage the flow of knowledge. Knowledge Management has been defined by Zou and Lim (2002) as the process undertaken by an organisation, with the buy in from all organisation levels, to create, capture, store and disseminate both tacit and explicit knowledge, through the use of IT tools, to aid organisational learning.

The maturity of a KM strategy is a critical component of its effectiveness (Maqsood, Finegan and Walker, 2006). Kruger and Snyman (2007) broke down KM maturity into 7 levels with the 6th level of maturity stipulating that an organisation was able to extend its KM strategy beyond its organisational boundaries. It is from this 6th stage of KM maturity that many academics, researches and KM practitioners believe that the real value of KM can be achieved as the potential value of KM is maximised due to the increased availability of knowledge to be entered into the organisations KM strategy (Orange, Burke and Boam, 2000; Maqsood, Finegan and Walker, 2006; Kruger and Johnson, 2009; Rezgui, Hopfe and Vorakulpipat, 2010). With the implementation of mature KM strategies by multiple organisations through several continuous improvement cycles the performance of the sector as a whole also has the potential to be uplifted (Orange, Burke and Boam, 2000). Kruger and Johnson (2009) found that the maturity of KM in South African construction organisations was being stunted as the KM strategies did not extend beyond organisational boundary. Therefore, KM in the sector is not as successful as it was hoped it would be.

The South African construction sector was identified as a key contributor to uplifting the nation following the fall of the Apartied in 1994 by the incoming government (Ofori, Hindlet and Hugo, 1996; McCutcheon, 2003). The sector previously had a proven track record of delivering successful projects and therefore Public Works projects were initiated to provide basic services and infrastructure nationwide, whilst also enabling socio-economic development and creating employment (Ofori, Hindlet and Hugo, 1996). The sector operates within a unique environment where the political landscape, wide range of languages spoken, races and cultures all play a part beyond the norm of the industry globally. By completing construction projects successfully the South African post 1994 construction sector had the opportunity to be a model of success to be emulated globally (Ofori, Hindlet and Hugo, 1996; Kruger and Johnson, 2009).

Recently the South African construction industry had their chance to show off its ability to complete large-scale infrastructure projects to a global audience with the building of the facilities for the 2010

FIFA World Cup. However, the various projects were all delivered late and experienced significant cost overruns (Baloyi and Bekker, 2011). This was typical of similar projects worldwide at the time and is still evident today (Baloyi and Bekker, 2011; Rivera *et al.*, 2017). The prevalent issues that caused the time and cost overruns were "material cost and price fluctuations", "incomplete drawings", "design changes" and "slow decision making" all which correlated with the global trends both in 2010 and still today (Baloyi and Bekker, 2011; Rivera *et al.*, 2017).

If the construction sector of South Africa has a proven track record of successful projects, why did these issues arise? Would lessons not have been learnt on past projects? The experience and knowledge gained should have been passed on to the latest project teams to mitigate the risk of these issues arising?

Through an examination of research by Latham, 1994; Ofori, Hindlet and Hugo, 1996; Orange, Burke and Boam, 2000; Cooper, Lyneis and Bryant, 2002; Kamara *et al.*, 2002; Schindler and Eppler, 2003; Senaratne and Malewana, 2011; Carrillo, Ruikar and Fuller, 2013 and Shokri-Ghasabeh and Chileshe, 2014 it was found that there are multiple barriers that exist globally that inhibit KM's implementation and growth. Issues such as the allocation of time for KM activities, the workplace culture and the proficiency of training given on the KM strategies implemented are a few examples of issues that have been found to impact on KM's performance. This research examined the barriers found in the literature and determined which could impact KM in South Africa, with emphasis placed on those that could limit an organisation to extend its KM activities beyond its borders.

Globally there has been some traction to address poor knowledge transfer and management of projects whereby long-term contractual relationships have been established between construction organisations to work together. Project knowledge sharing frameworks inclusive of multiple organisations have been set up to ensure that knowledge can freely flow between organisations and from one project to another. The cross-organisational learning approach (COLA) was developed by researchers in the UK who established a commitment between organisations to review and discuss recently completed projects to generate and disseminate knowledge to all contributing partners. These cross-organisation strategies are still relatively new, untested and require further research and investigation. However, the potential knowledge that can be extracted is extremely valuable and desirable to the organisations involved (Orange, Burke and Boam, 2000).

From the knowledge gained in the literature review a survey was developed which was issued to construction professionals throughout South Africa to gain their insight on the topic of KM and extending its activities beyond organisational borders. Through the statistical analysis completed on the data received this research was able to build on the findings of Kruger and Johnson (2009). The current opinion of the sector towards KM and extending KM activities beyond organisational borders was determined together with determining if there are barriers inhibiting these activities.

1.3 Problem Statement

The research of Kruger and Johnson (2009) raised the question of why South African organisations do not extend their KM practices beyond their organisational borders without being able to provide reasoning based on factual data. The literature has shown that by highlighting the barriers that inhibit KM's implementation the construction sector can overcome the challenges faced (Shokri-Ghasabeh and Chileshe, 2014). Therefore, to aid the South African construction sector improve upon its collective KM maturity, there is a need to examine the barriers faced by construction organisations when trying to extend KM activities beyond organisational borders. The current literature available on the topic shows that the fragmentation of the construction sector, a lack of knowledge and understanding of KM, the time allocated for KM activities, organisation culture, management support, contractual agreements and project continuity are some barriers and contributing factors restricting KM's implementation globally (Latham, 1994; Ofori, Hindlet and Hugo, 1996; Orange, Burke and Boam, 2000; Cooper, Lyneis and Bryant, 2002; Schindler and Eppler, 2003; Senaratne and Malewana, 2011; Carrillo, Ruikar and Fuller, 2013; Shokri-Ghasabeh and Chileshe, 2014). By determining if these barriers are prevalent in the South African construction sector, establishing if there is relationship between the barriers and an organisations KM maturity and finding out what the sectors current attitude is towards KM and extending KM activities beyond organisational borders this research will be able to provide insight to Project Managers and KM practitioners to refine their processes and strategies to improve their organisational KM performance.

1.4 Research Question

"What are the barriers and contributing factors inhibiting South African construction organisations from extending knowledge management activities beyond organisational borders?"

1.5 Research Aims

This study examines the barriers and contributing factors inhibiting organisations in the South African construction sector from extending KM activities beyond organisational borders.

1.6 Research Hypothesis

The null hypothesis to be tested in this research is that the following barriers and contributing factors have no statistically significant relationship to inhibiting South African construction organisations Knowledge Management activities from extending beyond organisational borders and the growth of their Knowledge Management maturity:

- Fragmentation in the construction sector.
- A lack of knowledge and understanding of KM and its activities and processes.
- A lack of time allocated for KM activities.
- An organisational culture not conducive to knowledge sharing.
- A lack of management support for KM.
- A contractual framework not conducive to knowledge sharing.
- A lack of continuity inhibiting KM

1.7 Research Objectives

The research objectives are to:

1. Determine a KM maturity level indicator for each respondent's organisation in the sample data utilised.
2. Determine if the following key barriers are prevalent:
   - The knowledge/training in KM provided by construction organisations is insufficient.
   - The culture within a construction organisation is not conducive to sharing knowledge.
   - The contractual framework of construction projects is not conducive to knowledge sharing.
   - Construction organisations are not incentivizing resources to interact and share knowledge, in addition to their normal project activities.
   - Not enough time is allocated for KM activities within construction projects.
3. Establish the factors behind these key barriers in order of prevalence and determine the severity of the impact these have on knowledge sharing and extending KM activities beyond organisational borders. The factors selected for analysis are as follows:
   - Time Allocated
   - KM Training
   - Management Support
   - Continuity
   - Contractual Framework
   - Job Security
   - Language Challenges
   - Temporary Multi-Organisation Coalition Challenges
4. Determine the current opinion towards KM and extending KM activities beyond organisational borders.

1.8 Research Method

This research will utilise a postpositivist quantitative approach as its research methodology. An online survey incorporating questions that provided insight on the research question posed was utilised to gather information for analysis. As the KM processes utilised by organisations are heavily reliant on human interactions it makes sense to conduct research on this area with a method that is tailored towards effectively gathering and understanding data for social problems. The online survey provided a data set of information that was analysed by the researcher (Creswell, 2014).

The online survey was distributed to professionals in the South African construction industry. The professionals contacted were from the Association of Construction Project Managers and contacts that the author of the research had made in the construction sector.

From the data collected a rich picture of the South African construction sector was developed with regards to the barriers inhibiting KM activities. The research analysed the data and then compared the findings to the available literature. Finally, the research provided constructive feedback to the construction sector on how to overcome the barriers found.

1.9 Scope of the Study

This research is limited to exploring the barriers that inhibit KM activities, with emphasis placed on those thought to inhibit these activities from extending beyond organisation borders. The research will only look at barriers that affect KM activities in construction sector, focussing in on the teams that drive a construction projects delivery.

1.10 Significance of the Study

This research has the potential to drive change within the construction sector. By outlining the factors that are inhibiting KM from extending beyond organisation borders this research has the potential to enlighten KM practitioners to address their organisations KM strategies and implement change. Through this change the KM maturity can be improved upon and ultimately the performance of the construction sector can improve.

1.11 Structure of the Research Report

The report will be structured as follows:

1. **Introduction** – A brief outline of the research topic is discussed followed by a description of the research problem and the objectives that the research will seek to achieve.
2. **Literature Review** – a critical review of the literature available relevant to the research topic is completed in order to establish the base of knowledge existing on the topic.
3. **Research Methodology** – this chapter details how the research was carried out and discusses why the chosen methodology was considered best to this research topic.
4. **Data Presentation, Analysis and Discussion** – this chapter analyses all the data gathered from the survey. The results of the research are interpreted and discussed in detail.
5. **Conclusion and Recommendations** – in this chapter conclusions are drawn based on the findings of the research, relating back to the research question posed. Recommendations for the construction sector and future studies are made.
- **References** – a list of all sources of all information utilised in the research was developed.

2. Literature Review

2.1 Introduction

This chapter presents the evaluation and analysis of the available literature on the topic of knowledge management and how its implementation has been limited in the construction sector. Through the examination of the literature a base of knowledge was established, and the foundation of the research problem was established.

2.2 Overview of the Construction Sector and Projects

The construction sector globally is still highly informal whereby resources typically rely heavily on their experience gained, skills developed and do not generally follow analytical methods or textbooks. Knowledge is typically gained through years of on-the-job training which is shared predominantly via word of mouth, if at all (Carrillo and Chinowsky, 2005; Maqsood, Finegan and Walker, 2006). The construction sector is knowledge intensive due to the nature of tasks undertaken by the resources to produce ever more complex products. This results in new knowledge regularly being created which could be captured, however the informal trait of the construction sector acts as an inhibitor to capturing this knowledge (Senaratne and Sexton, 2009).

The construction sector is a project-based environment with organisations working to deliver the desired product to the projects sponsors. These projects are each a temporary endeavour, unique from all other projects due to the vast array of factors that make up a project. However, whilst every project is different, they all share multiple commonalities, processes and activities which are repeated throughout a projects life cycle (Cooper, Lyneis and Bryant, 2002; Carrillo, 2005).

One common factor of every construction project is that a team of resources is needed to complete the various project tasks to deliver the project. The resources needed to complete a construction project rarely all come from one organisation and therefore each project will typically form a new temporary multi-organisation (TMO) coalition for the duration of the project (Bakker *et al.*, 2011; Fellows and Liu, 2012). The number of organisations that form this TMO is in part dependant on the number of specialists needed, and as construction projects are increasing in complexity, the number of specialists is increasing. The TMO will include resources that could be assigned to one, multiple or all phases of the project. Further, in the modern global project environment, the TMO can be separated geographically and need to interact via IT platforms, resulting in "virtual" TMOs being formed (Fellows and Liu, 2012). As the success of a construction project is reliant on the effective completion of numerous activities, by multiple different resources, from different organisations, in parallel or sequentially, the performance of the project is critically dependant on the effectiveness of the interactions between organisations and how freely knowledge flows beyond each organisation's boundary. The dynamic nature of the resources entering and exiting the project, the need for numerous specialists, geographical separation and the challenges of knowledge flow between organisations not only impacts the performance of the project but further makes it difficult to ensure that the knowledge accumulated by the various resources during the project is effectively captured and disseminated to the whole TMO (Fellows and Liu, 2012). These factors have contributed towards the label of "fragmentation" in the construction sector which has become a "tagline" to explain away many of the problems faced by the construction industry (Cooper, Lyneis and Bryant, 2002; Kamara *et al.*, 2002; Carrillo, 2005; Bhargav and Lauri, 2009; Fellows and Liu, 2012).

The performance of construction projects globally has been under scrutiny for decades as many projects often experience delays, exceed their budgets or do not meet the quality standards required, leading to dissatisfied project sponsors (Senaratne and Sexton, 2009). Research conducted by Egan (1998) found that more than a third of all project sponsors were dissatisfied with the delivery of their construction projects. A recent study conducted by the Construction Industry Institute found that only 2.5% of projects were defined as successful with billions of dollars being wasted due to inefficient practices (CII, 2015). History is littered with large mega projects that are famous due to the extent that their budgets overran with the Channel Tunnel between England and France a great example. The final budget of £4650 million was 80% over the initial estimate of £2600 million. Research by Rivera *et al.* (2017) has shown that there has been marginal improvements in cost overruns globally. However, rework on projects was found to still be responsible for up to 6% of the projects cost. This was found predominantly due to variation orders and design errors. Interestingly, contrary to popular belief, recent research has found there is no discrepancy between the poor project performance of developing and developed countries (Rivera *et al.*, 2017).

The performance of a project will be impacted by a wide range of factors. Recent research by Durdyev and Hosseini (2019) completed a review on research papers from 1985 to 2018 that examined the causes of delays on construction project in order to determine a comprehensive list of the most common factors in both developed and developing countries. Their research identified the following factors in Table 1 below to be the 10 most common causes of delays on construction projects in both developing and developed countries.

Table 1: Causes of Delays on Construction Projects – Source Durdyev and Hosseini (2019)

#	Factors
1	Climate conditions
2	Poor communication
3	Lack of coordination and conflict between stakeholders
4	Ineffective planning
5	Material shortages
6	Financial issues
7	Payment delays
8	Equipment shortages
9	Lack of experience/qualifications/competence
10	Labour shortages and poor site management

With the identification of the most common causes of delays on construction projects the sector should be able to tackle and address each in order to improve its collective performance. However, as the causes are largely interconnected and influence each other it would be an oversimplification to try and address each cause individually.

2.3 Concept of Knowledge Management

With the rise of globalisation in the 1980s, organisations faced competition on a new scale. To remain competitive in this new business environment organisations found that it was critical to safeguard their collective knowledge and to manage it as a business asset (Wiig, 1997; Rezgui, Hopfe and Vorakulpipat, 2010). It is an organisations knowledge that enables businesses activities to be completed and knowledge can be viewed as the ultimate driver of the organisations profitability (Terzieva, 2014). Early research on the topic of explicit knowledge management showed that there was an appreciation for the importance of managing knowledge by business leaders, however there was little consensus on how to achieve effective KM. As each business sector had its own management style, operational procedures and organisational goals, the KM activities developed differed for each sector (Terzieva, 2014). It was predominantly in the IT sector that KM gained early traction and rose to prominence (Wiig, 1997).

KM as a business strategy was developed in order to aid the organisation improve its performance by extracting the maximum value from the knowledge that is held by the organisation. As KM developed it has become a broad, multi-disciplinary strategy that can now influence every aspect of an organisation (Wiig, 1997; Rezgui, Hopfe and Vorakulpipat, 2010).

As KM was developed in parallel by multiple industries to suit different business goals there have been multiple definitions developed through the years. KM can be viewed as the identification and capturing of knowledge created within an organisation that has potential value (Zou and Lim, 2002). This knowledge can then be managed effectively to be stored and disseminated throughout an organisation to be utilised as an intellectual asset. Research has shown that through knowledge that problem solving activities are improved, the time taken on tasks is reduced, wasted activities are eliminated and innovation is promoted (Skyrme and Amidon, 1997; Kamara *et al.*, 2002; Zou and Lim, 2002; Shokri-Ghasabeh and Chileshe, 2014).

2.3.1 Knowledge

As KM is reliant on the transferring of knowledge it is important to first have a sound understanding of the concept of what "knowledge" is. Orange *et al.* (2000) claimed that knowledge is a product of an individual learning from, interpreting and/or processing the information to which they are subjected to on a daily basis. Whereby information is described as the expression of knowledge, which is capable of being stored, accessed and communicated.

Knowledge can be explicit, whereby it is written down and stored in a documented form. Alternatively, knowledge can be tacit, whereby it is stored within an individual's mind, based on their personal experience. Tacit knowledge shared socially and is the dominant method of knowledge sharing in construction projects, however tacit knowledge can be complex and therefore difficult to communicate to others (Senaratne and Sexton, 2008; Carrillo, Ruikar and Fuller, 2013).

In summary, knowledge can be seen to be a highly individualised item whereby each employee will have a unique viewpoint of the same data. This knowledge held by each employee is then tacitly stored individually and with each passing moment can be modified based on new experiences and inaccurately recalled at a later stage or lost entirely (Senaratne and Sexton, 2008; Carrillo, Ruikar and Fuller, 2013).

This presents a challenge when developing a management system methodology that aims to accurately capture knowledge generated by employees when it is possible that the employees will each create different knowledge when presented with the same information. KM practitioners therefore had to be cognisant of this when developing the methods used to capture the knowledge of the lessons learned.

Construction projects have been established to be unique endeavours which implies that on each project unique new tacit knowledge will be created (Bakker *et al.*, 2011). It is the role of the KM activities to then capture this knowledge in each project to ensure that it is not lost. This is often difficult to complete accurately and is generally time intensive (Bakker *et al.*, 2011). The KM activities utilised therefore need to be developed with the awareness of these challenges in mind to ensure their successful completion (Bakker *et al.*, 2011; Terzieva, 2014). An organisations knowledge is created by its employees through their individual knowledge that is then disseminated formally through the KM strategy or informally through personal interactions. The extent of knowledge sharing within an organisation is heavily influenced by the attitude of its employees towards sharing their own individual knowledge (Peihua and Fung, 2013).

2.3.2 Knowledge Management in Practice

As discussed, KM's implementation is able to be tailored to suit organisational goals. Therefore, there are numerous methodologies that have been developed and utilised. For example, popular methodologies in use are Project Audits, Post Project Reviews (PPR), After Action Reviews (AAR), Lesson Learned documents (LL) and Learning Histories.

These methodologies fall into one of the following basic approaches or are a combination of the two. The first approach is supply driven whereby the KM strategy seeks to enable the flow of knowledge throughout the organisation. This is facilitated predominantly using an ICT tool that captures, codifies and disseminates organisational knowledge. The supply driven approach is typically mechanistic and is proficient at capturing explicit knowledge (Kamara *et al.*, 2002; Maqsood, Finegan and Walker, 2006). The second approach is demand driven whereby the KM strategy facilitates and encourages its users to share their experiences, successes and failures to their co-workers. This is a softer, more organic, approach that typically utilises storytelling and communities of practice to enable sharing of knowledge (Kamara *et al.*, 2002; Maqsood, Finegan and Walker, 2006).

Project-based organisation such as those typical of the construction sector, for example, have been known to make use of a PPR to capture the knowledge generated throughout the project (Shokri-Ghasabeh and Chileshe, 2014). A PPR gathers all the roles players involved in a project to discuss the project's successes and failures to capture the valuable knowledge that has been generated. Through an effective KM strategy each organisation involved this knowledge can then be passed on to future project teams to improve their performance and project delivery (Shokri-Ghasabeh and Chileshe, 2014).

KM as a strategy or methodology has largely remain informal until the advancements in technology which has enabled the easy codification, storage and transferring of large amounts of knowledge on platforms such as intranets (Rezgui, Hopfe and Vorakulpipat, 2010). The practice of capturing knowledge generated by a project has been championed by academic powerhouses such as the Project Management Institute (PMI) who have included audit/review activities into their project management book of knowledge, the PMBOK (*A Guide to the Project Management Body of Knowledge (PMBOK Guide)*. 4th Editio, 2008). Industry leading organisations such as Swiss bank UBS, manufacturing giant Schneider and technology companies like IBM have all adopted KM strategies into their organisations due to its massive potential value (Schindler and Eppler, 2003).

2.3.3 Knowledge Management in the Global Construction Sector

Egbu, Sturges and Bates (1999) recognised that KM would be a strategy suited to the construction sector to achieve improved project performance. Numerous research papers have since confirmed Egbu, Sturges and Bates (1999) claim by showing how KM has been able to effectively drive innovation, improve performance and reduce costs if implemented correctly (Kamara *et al.*, 2002; Rezgui, Hopfe and Vorakulpipat, 2010). KM's success has seen its implementation in many of the leading organisations in the global construction sector such as Amec, ARUP, Balfour, Taylor Woodrow, Turner & Townend and Wates Construction (Carrillo and Chinowsky, 2005).

2.3.4 Knowledge Management Maturity

Each construction organisation utilising KM would have developed a strategy that would suit its own business needs, company culture and operating structure. Whilst the success of each organisations KM strategy is the result of many complex factors, it is heavily influenced by the maturity of the KM activities and processes established (Rezgui, Hopfe and Vorakulpipat, 2010). Kruger and Snyman (2007) argue that the maturity of a KM strategy is directly related to the speed of an organisation's innovation cycle.

In order to then evaluate the maturity of an organisations KM strategy it needs to be baselined from a theoretical model. There have been many KM maturity models developed, however most have been influenced by the ICT sector, which emphasised technology over the softer aspects of KM strategies. In response to the need for a holistic KM maturity model Kruger and Snyman (2007) developed the baseline in Table 2 below:

Table 2: Knowledge Management Maturity Levels – Source Kruger and Snyman (2007)

Level	Description
1	The organisation must have an information management process/strategy that utilises an ICT platform that can store information that is collected and codified by the organisation.
2	The organisation must have realized the importance of knowledge management as a formalised strategy that has been communicated throughout the organisation.
3	The organisation has committed to undertake KM activities with senior management driving its implementation. The ICT system has developed beyond only enabling the flow of information to now aid decision making and strategies.
4	The organisation has developed strategies specifically focussed on exploiting its knowledge as an asset.
5	The organisation's KM strategy has developed to the point where the KM processes are streamlined and improved upon by the KM strategy itself. The organisations culture has become conducive to knowledge sharing within its ranks.
6	The organisation has developed successful partnerships with all stakeholders within its value chain that allow for knowledge sharing beyond organisational borders. These partnerships open each organisation to wider base of knowledge which has potential value when incorporated into their KM strategies.
7	The KM strategy has developed to the position where employees are driving the evolution of KM and proposing new innovative KM strategies to be implemented in the organisations near future.

The Knowledge Management model developed by Kruger and Snyman (2007) differentiates itself from other KM maturity models by highlighting that the maturity of a KM strategy is not dependent on the sophistication of the ICT system utilised. The maturity of a KM strategy is rather dependent on the managerial strategies implemented, the cultural development of the organisation, the activities/processes undertaken by employees and the type of social interactions that are facilitated in the workplace (Kruger and Johnson, 2009).

An ICT system is able to store and disseminate information, but this does not equate to knowledge sharing. Individuals create knowledge through understanding which relies on how the information is presented, communicated and embedded though the socialisation of the information received. Numerous "soft" factors need to be taken into consideration to facilitate this socialisation such as the attitude of employee's towards knowledge creation and the strength of the relationship between the employees interacting (Jacky *et al.*, 1999).

Therefore, the successful implementation of KM is a complex challenge to take on, with the adoption of an ICT system alone not being sufficient. KM has been found to deliver quantifiable value to any organisation once the strategies implemented begin to facilitate a culture of continuous improvement – based on Kruger and Snyman's (2007) maturity model this would be once a maturity level of 4 is reached. Innovation, efficiency and productivity within an organisation have seen improvement once this level of KM maturity has been achieved (Maqsood, Finegan and Walker, 2006; Rezgui, Hopfe and Vorakulpipat,

2010). Research by Kruger and Johnson (2009) confirmed a link between organisational performance and the maturity of the KM strategy implemented.

2.4 The South African Construction Sector

Despite emerging from apartheid over 25 years ago South Africa still faces large scale inequality, poverty and unemployment. The construction sector was identified by the newly elected government in 1994 to be a key vehicle to redress the errors of the past administration and uplift the socio-economic state of the country (Ofori, Hindlet and Hugo, 1996) . Historically South Africa has a proven track record of constructing and managing large-scale built environment projects from bridges, dams, airports and skyscrapers both locally and throughout Africa. South Africa has a large unskilled workforce seeking employment and many tertiary education institutes that produce the skilled technical resources needed to manage and implement the public amenities, housing and the associated services infrastructure build projects needed (Ofori, Hindlet and Hugo, 1996).

However, the large-scale public works programme implemented post-apartheid have broadly been labelled a failure with basic infrastructure services still not available nationwide and the construction industry stagnating (Ofori, Hindlet and Hugo, 1996). The construction sector faces unique challenges to deliver these projects due to numerous complex issues such as continued political unrest, widespread corruption, a broad lack of education and technical skills, contrasting managerial and work styles influenced by the various cultures and ethnicities, a geographical dispersion of resources throughout the country and a lack of financial investment needed (McCutcheon, 2003).

Recently the construction sector had a spotlight upon it with the large-scale infrastructure projects completed for the 2010 Soccer World Cup (SWC). This was an opportunity for the local sector to shine and regain the confidence of the South African citizens following the failures of the early public works projects. However, all 10 stadiums were completed with cost overruns and/or late (Baloyi and Bekker, 2011). Research completed by Baloyi and Bekker (2011) found that the 2010 SWC stadium projects were hampered by late payments, increases in material costs, environmental factors, resource fluctuations, poor contract management. inaccurate estimates and corruption.

Despite the sector's poor delivery thus far, the construction sector is a critical part of South Africa's economy and is a significant contributor to its growth (CIDB, 2012). Typical of the construction sector globally, the makeup of the sector is similar in South Africa, where a few large organisations dominate the market with a growing proportion of SME's (Thwala and Phaladi, 2009). These SME's can contribute to a substantial portion of a countries GPD whilst also being a large employer of unskilled labour. The South African government has identified this growing segment of SME's as an area to promote their strategy of BBBEE to redress the status quo of the monopoly present in the sector. SME's can also operate on the fringes and in remote areas that large contractors are not typically willing to focus on. This is important in the South African context as the most vulnerable citizens in need of basic services are located in these remote areas (Thwala and Phaladi, 2009).

To assist SME's locally to grow the South African government, through the Department of Public Works (1997) formulated the Integrated Emerging Contractor Development Model (IECDM), which is based on the Emerging Contractor Development Model (ECDM) developed by CSIR. This was a best practice tool aimed to assist emerging contractors grow in the industry by focussing on developing a business plan, project management basics, Total Quality Management (TQM) principals & strategies and promoting nationally accredited qualifications (Thwala and Phaladi, 2009).

The project management and TQM principles touched on many KM processes such as conducting stage gate reviews and continuous improvement loops. However, thus far the strategy has not been as successful as hoped with a poor participation rate (Thwala and Phaladi, 2009). The need remains to provide support to the emerging SME's, with a special focus on supporting females in the sector, remote and vulnerable areas, increasing access to training, funding and business opportunities (McCutcheon, 2003; Thwala and Phaladi, 2009).

South Africa's economic climate is slowly emerging from a decade of economic weakness where it fell from 44th in 2007 to 67th in 2018 on the Global Competitiveness index. In 2019 South Africa recorded a GPD growth of only 1.5%, which is expected to improve only moderately to 2.1% by 2021. South Africa's economy faces many challenges and issues such as the risk of power outages due to the poor state of Eskom (its state-run power generation and supply company), which has knocked investor confidence. Whilst the government has earmarked R500 billion towards infrastructure projects over the next 3 years, activity and employment in the construction sector has weakened due to declining investment Index (National Treasury: Republic of South Africa, 2019). Whilst the poor economic environment is no doubt a contributing factor to the poor performance of the construction sector, poor management practices have been identified to have contributed to the failure of many organisations (Thwala and Phaladi, 2009).

The IECDM strategies implementation along with research completed by Kruger and Johnson (2009) shows that there has been an uptake in KM practices in the South African construction sector. The harsh South African environment described in which the construction sector operates has contributed towards local organisations having to look at strategies such as KM to remain competitive. Growth of KM has been both in its uptake and with the maturity level being achieved. However, this has predominantly been seen in larger organisations who have the resources and funding available to implement the required KM strategies (Kruger and Johnson, 2009c).

South Africa had the potential to be as success story and a model for the world to follow. The author of this research has witnessed how South Africa has multi-organisation project teams who communicated via many different languages, often using their 2nd or 3rd languages. There are clashes of different cultures which could influence communication, management styles and the sector has organisations with varying levels of KM maturity (Kruger and Johnson, 2009b). Further, the struggles of the past, have left an inherent distrust between those previously disadvantaged and those who were empowered. The need to redress the inequalities of the past have seen the introduction of schemes such as BBBEE, Affirmative Action, Quotas and the proposal of redistribution of land without compensation, which adds an extra layer of complexity to interactions and especially conflict management between project stake-holders (Kruger and Johnson, 2009c). Despite the challenges faced, the South African project teams still successfully complete construction projects in some of the harshest natural and economic environ-ments globally. In the modern world of globalisation and the growing entanglement between different cultures, the global projects of the future will need to be able to effectively cope with complex diversity, a phenomenon evident in most construction projects in South Africa (Kruger and Johnson, 2009a).

2.5 Barriers to Extending Knowledge Management Beyond Organisational Borders in the Construction Sector

In the following section of this literature review examines the barriers found in the literature that have been found to inhibit KM activities from extending beyond organisational borders. Where barriers were found to consist of multiple factors each factor was detailed and discussed to provide clarity on its impact on extending KM activities beyond organisational borders.

2.5.1 Fragmentation of the Construction Sector

The fragmentation of the construction sector presents many challenges that have the potential to inhibit knowledge transfer and KM activities. The various factors of fragmentation that inhibit KM activities from extending beyond organisational borders are discussed as follows:

2.5.1.1 Temporary Multi-Organisation Coalition (TMO's) Challenges

The project resources that make up a TMO enter in and out of the challenging and stressful environment of the project to work together to complete critical tasks. As the resources of a TMO are dynamic it can be difficult to ensure that the correct data is collected from the right individuals at the right time. For example, if a Post Project Review (PPR) is utilised it is possible that member of the TMO who were only involved in the initial stages of a project will not be present and therefore the knowledge they hold will not be captured and disseminated to the full TMO. The fluid nature of resources in and out of the project also impacts the TMO's ability to determine the root cause of a problem as the key knowledge could be held by a resource that has left the TMO already. The challenge faced then to organise and ensure that knowledge is extracted from project team members effectively is an inhibitor not only KM activities but also to extending these activities beyond each organisation's borders within the TMO.

2.5.1.2 Broken Feedback Loops

Senaratne and Malewana (2011) found that the construction sector typically has standard operating procedures that do not have completed feedback loops. These broken feedback loops inhibit knowledge transfer (Senaratne and Malewana, 2011) and would therefore act as a barrier to extending KM activi-ties beyond organisational borders.

2.5.1.3 Resource Mobility

Key resources within the world's largest and most complex projects will typically only be involved in a handful of projects before moving out of their project role or retiring. Staff turnover is also of concern in the modern business environment where resources have been seen to be more fluid and willing to move between organisations, potentially taking unshared knowledge with them (Schindler and Eppler, 2003).

2.5.2 Knowledge and Understanding of KM Activities and Processes

KM processes, procedures and activities can often be time intensive, laborious, IT intense, complicated and/or vague to completed. This can inhibit KM's effectiveness unless employees have a clear understanding of what KM is, why it is being implemented, its potential value to the organisation and how to complete the various activities. It is critical to ensure that the resources allocated to these tasks receive the required training, motivation and incentivisation to complete the tasks effectively. Without the effective participation and buy-in of all required resources the KM strategy will not maximise the value extracted from the available knowledge (Schindler and Eppler, 2003; Shokri-Ghasabeh and Chileshe, 2014). Critical to the training received in KM employees must understand the suitability of KM as a management tool to use. As defined, a project is a temporary endeavour, undertaken to create a unique product and/or service. As KM is reliant on utilising knowledge generated in past experiences to influence future endeavours would it be suitable in a project intense environment? Whilst the product, system, environment, stakeholders and complexity will almost certainly vary from project to project there are similarities throughout the project life cycle that can enable cross-project learning. Problem solving techniques, contractual negotiations, conflict resolution strategies, established feedback loops, project processes, compliance and regulatory steps are all similarities within the construction project sphere where it is possible for KM to facilitate continuous improvement (Cooper, Lyneis and Bryant, 2002; Kamara *et al.*, 2002). With insufficient training in KM the effectiveness of the KM activities both internally and extending beyond organisation boundaries will be inhibited.

2.5.3 Time Allocated to KM Activities

The need for time to be set aside for KM activities to take place must also be considered. Projects in the construction sector are fast paced, activity intense and highly stressful for the resources involved. Organisations tend to ensure that perceived "mission critical" tasks are prioritised over the softer "nice to have" tasks such as setting aside time for reviews and/or project audits (Schindler and Eppler, 2003; Shokri-Ghasabeh and Chileshe, 2014). Without enough time allocated to all KM activities, both internally and when extending beyond organisational borders, the effectiveness of the KM strategy is reduced.

2.5.4 Organisational Culture

The culture of the organisation must also be considered as an inhibitor to knowledge transfer and KM activities. Not wanting to admit faults, internal competition and protecting job security can limit knowledge transfer between employees. These factors have the potential to reduce the effectiveness of a KM strategy implemented and each factor is discussed in detail as follows:

2.5.4.1   Admitting Fault

The essence of KM is for reflection on past mistakes to prevent future errors. The culture of most organisations has been found to be that errors and mistakes are often not spoken about as employees fear persecution and/or retribution (Schindler and Eppler, 2003; Carrillo, Ruikar and Fuller, 2013).

2.5.4.2   Competition within Organisations

Within organisations it is also possible to have different departments and/or satellite offices who are reluctant to work together. Competition between parts of an organisation can limits the transfer of knowledge. This phenomenon ties into the idea that "knowledge is power" and should be protected (Carrillo, Ruikar and Fuller, 2013).

2.5.4.3   Job Security

Building in the idea that "knowledge is power" there are also instances where employees are reluctant to share knowledge in order to protect their employment. In South Africa, with the implementation of structures such as BBBEE, Affirmative Action and quotas, there is the potential that those who hold knowledge are unwilling to share to protect their employment (Ofori, Hindlet and Hugo, 1996).

2.5.5 Management Support

The role that management plays in enforcing KM's implementation must also be highlighted. Without the support and drive from the top down to carry out KM activities, both internally and beyond organisational borders, the effectiveness of the KM strategy was found to be negatively impacted (Schindler and Eppler, 2003).

2.5.6 Contractual Agreements

The contractual agreements that bind the TMO together also has the potential to hinder the transfer of knowledge throughout the TMO during the project's life cycle (Carrillo, Ruikar and Fuller, 2013). The competitive dynamic between organisations within the sector must also be considered as organisations can be reluctant to share knowledge to protect their competitive advantage and intellectual property, which extends into the culture of the construction sector. Research by Latham (1994) found that the UK construction sector displayed a culture where organisations developed defensive relationships with their TMO partners. Organisation actively sought to find faults in each other's work and litigation was prevalent when resolving conflicts. This highlights that the admittance of faults and errors during and/or after a project potentially opens an organisation up to legal persecution (Schindler and Eppler, 2003; Carrillo, Ruikar and Fuller, 2013). If the contractual agreements are not established to allow for open and honest project reviews and enable knowledge transfer between organisations within the TMO the effectiveness of KM activities extending beyond organisational borders is inhibited.

2.5.7 Continuity

The challenges faced to implement KM effectively in the construction sector can to a large extent be tied back to a common root issue of a lack of continuity in the project teams that deliver projects. Research by Orange, Burke and Boam (2000) found that not only did the lack of continuity inhibit KM's implementation but the associated organisational learning opportunities. Research into continuity by Egan (1998) highlights how critical this is in a team's composition. He stated that:

"The repeated selection of new teams in our view inhibits learning, innovation and the development of skilled and experienced teams."

And:

"A team that does not stay together has no learning capability and no chance of making the incremental improvements that improve efficiency over the long term."

These statements by Egan (1998) are damning of KM's implementation in the construction sector where the continuity of project teams is so rare. In response to Egan the question posed by the Author of this research is then:

"Is it possible that the construction industry can change how project TMO's are created? Is it possible to change the status quo and is it realistic to consider a scenario where teams are kept constant over multiple projects?"

Perhaps the more realistic path to follow is to seek to address the culture within the sector to enable knowledge transfer instead of trying to restructure the whole system. By promoting the culture of collaboration and the spirit of sharing knowledge there is a possibility of more construction organisations achieving mature KM strategies despite Egan's misgivings on TMO's.

2.5.8 Summary of Barriers to Extending Knowledge Management Beyond Organisational Borders in the Construction Sector

KM's implementation in the construction sector is potentially hindered by multiple barriers which impacts its effectiveness. Knowledge has been found globally to be remaining within the project TMO members and it is not being disseminated both throughout and beyond the TMO (Rezgui, Hopfe and Vorakulpipat, 2010; Carrillo, Ruikar and Fuller, 2013). This level of KM correlates with the early stages of the KM maturity model developed by Kruger and Snyman (2007) and therefore, organisations are not unlocking the full potential value of the knowledge available to them (Maqsood, Finegan and Walker, 2006).

Recent studies have shown that 35% of companies worldwide are using KM with a satisfaction rating of 70%. Therefore, despite the populism that KM has received in the academic field it is still not being universally implemented and further nearly a third of users are not satisfied with its impact on their organisation (Kruger and Johnson, 2009a).

2.6 Cross-Organisational Learning Agreements, Partnerships and Relationships

Globally there has been traction to address the barriers encountered in implementing KM activities and extending these beyond organisational borders. KM strategies are being revised, updated and developed to improve collaboration between employees and organisations and to drive knowledge sharing.

Long-term contractual relationships have been established between construction organisations to work together on repeated projects (Egbu, Sturges and Bates, 1999). Project sponsors are also including the requirement to include continuous improvement strategies and/or KM practices into their contractual terms (Carrillo, 2005; Carrillo, Ruikar and Fuller, 2013). Project Managers are pushing for more regular project meetings with the full TMO which can be facilitated globally with improved virtual meeting facilities (Egbu, Sturges and Bates, 1999).

Project knowledge sharing frameworks inclusive of multiple organisations have been set up to ensure that knowledge can freely flow between organisations and from one project to another. The "Cross-Organisational Learning Approach" (COLA) was developed by researchers in the UK who established a commitment between organisations to review and discuss recently completed projects to generate and disseminate knowledge to all contributing partners (Orange, Burke and Boam, 2000).

The partnerships and collaborations that are formed through these contracts have been found to improve knowledge transfer, KM practices and therefore multiple factors of project delivery (Bakker *et al.*, 2011). Feedback loops are closed when the project team maintains continuity rather than knowledge being lost (Franco, Cushman and Rosenhead, 2004). Essential project soft skills such as communication and the ability to share knowledge have been found to be improved when exposed to interactions with TMO members from a variety of cultures, perspectives and disciplines which are found in collaborative environments (Senaratne and Malewana, 2011). Jacky *et al.* (1999) established that there was a relationship between an organisation's innovation capability and their effective use of its knowledge, created by cross-functional, inter-organisational and inter-disciplinary teams. Further partnerships have been found to be an effective stagey to secure repeated business (Franco, Cushman and Rosenhead, 2004).

Despite the established benefits of a mature KM strategy for an organisation Kruger and Johnson (2009) found that South African organisations are not extending their KM practices beyond their own organisational boundaries. They posed that local organisations had a perception that extending KM beyond organisation borders would impact their organisation negatively by reducing their competitiveness in the sector. This perception ties back to the boundary paradox of how to have open organisation boundaries that allow information to flow both in and out, whilst still being able to protect the intellectual property of the organisation that gives it their competitive advantage (Egbu, Sturges and Bates, 1999; Fellows and Liu, 2012). The counter argument to this is then that as knowledge is changing so rapidly in the modern business environment that it may be short-sighted to set up boundaries, which protect IP, rather than collaborating to increase the mutually beneficial knowledge created (Hardy, Phillips and Lawrence, 2003).

Further, Kruger and Johnson (2009) found that there is little motivation and drive in the local sector to share knowledge. Employees have been found to be resistant to learning from others (Schindler and Eppler, 2003). The findings of their research indicate that there is a disconnect between the implementation of KM locally and what international research is pushing for through initiatives such as COLA.

2.7 Summary of the Chapter

This literature review examined the literature available on knowledge management, its implementation in the construction sector with a detailed look at the South African sector. This was completed to gain insight into the topic of knowledge management and the challenges being faced to its effective implementation in the construction sector. The research of Kruger and Johnson (2009) raised the question of why South African organisations do not extend their KM practices beyond their organisational borders without being able to provide an evidence based conclusion. The literature has shown that by highlighting the barriers that inhibit KM's implementation the construction sector can overcome the challenges faced. Therefore, to aid the South African construction sector to improve upon its collective KM maturity, there is a need to examine the barriers faced by construction organisations when trying to extend KM activities beyond organisational borders. By determining these factors this research will be able to provide insight to Project Managers and KM practitioners to refine their organisational processes and strategies to improve KM performance.

3. Research Methodology

3.1 Introduction

This research report examined barriers inhibiting organisations in the South African construction sector from extending their KM activities beyond organisational borders.

The propositions to be tested in this research are that the following inhibitors: a lack of knowledge/training in KM; a culture of protecting knowledge over sharing within organisations; contractual frameworks are not conducive to knowledge sharing; a lack of incentive for team members to interact beyond project specific activities and a lack of time is allocated for KM activities with the project are prevalent in construction project TMO's and construction organisations in South Africa, which are then acting as barriers to extending KM beyond organisational borders.

3.2 Research Approach

This research employed a postpositivist quantitative approach as its research methodology. Quantitative method involves the process of collecting, analysing, interpreting and drawing conclusions based on the objectives of the research. This research methodology was suited to this research as the objective of the research is to disprove the null hypothesis posed by determining if there is a statistically significant relationship between barriers identified in the literature and their impact on KM in the South African construction sector. This research will utilise a deterministic philosophy in which causes determine effects or outcomes. However, as knowledge is conjectural, the evidence established in this research is imperfect and fallible. The research null hypothesis presented are not able to be disproved, but rather the evidence developed will assist in supporting the rejection of the null hypothesises (Creswell, 2014).

3.3 Population of the Study

The population to be researched was ringfenced to include all construction sector professionals who regularly participated in the project team that delivered construction projects such as: Project Managers; Contractors; Engineering Consultants; Specialists; Quantity Surveyors; Architects; Property Developers and Resource Suppliers. The personal of the professional team was selected to be the population of the research as these employees who are exposed to project knowledge and are able to feed this knowledge back into their organisations KM strategy.

According to data from the CIDB (2019) in the 3rd quarter of 2019 South Africa reported a total labour force of 23.1 million with an employment rate of 29.1%. The Construction sector accounted for 8% of formal employment with a total 611 315 jobs. The total employment of consultants within the construction sector was found to be 94 200 employees who would therefore represent the total population identified for this research.

The sample frame of the research consisted of various members of the South African construction sector who could be contacted via their membership to professional bodies, associations and councils.

Organisations such as the South African chapter of the Project Management Institute (PMI), the Association of Construction Project Managers (ACPM), Project Management South Africa (PMSA) and the South African Council for the Project and Construction Management Profession (SACPMP) were contacted to participate in the research. However, due to various organisational policies, only the PMI (partially - 18 members) and the ACPM (325 members) shared their members contact details. Additionally, a sample frame of 156 personal contacts of the researcher, who are fellow professionals in the construction sector, were included in the sample frame.

3.4 Sampling Technique and Sample Size

The sample frame for this research consisted of a total of 499 contacts, which represented 0.53% of the population available. The online survey was distributed to all contacts in the sample frame with the survey initially sent out in September 2019, remaining open for responses until January 2020. The data collected for this research was cross-sectional and only collected during this period. The survey had a 38% completion rate and would typically take 17 minutes for a respondent to complete. A total of 181 responses were collected from the sample frame approached.

Through an examination of the responses multiple entries were discounted from the analysis due to the following issues:

- 11 data sets did not indicate that permission was granted to use the information shared.
- 5 data sets indicated that their organisation did not have a presence in the South African construction sector.
- 72 data sets failed to provide feedback beyond section 1 of the survey.

The remaining 93 data sets were screened for potential errors by examining the maximum and minimum values supplied and missing information was checked to ensure the validity of the data set was not compromised. The result of this check was that no further data sets were discounted for analysis yielding 93 validated data sets for analysis.

3.5 Method of Data Collection

The initial step taken in this research was to examine the available literature on knowledge management and its implementation in the construction sector both within South Africa and globally. From this base of knowledge, a survey was developed to ascertain the perspectives of South African project TMO members in relation to KM barriers and extending KM activities beyond organisational borders. The online survey was developed to gain insight of both the employee and that of the organisation that they represent.

The research utilised an online survey incorporating questions that provided insight on the research questions posed to gather information. As the KM processes utilised by organisations are heavily reliant on human interactions it was applicable to conduct research on this topic with a method that is tailored towards effectively gathering and understanding data for social problems (Creswell, 2014).

The questionnaire, as found in Addendum A, was developed and distributed utilising the online tool: SurveyMonkey.com. The online survey consisted of the following sections:

- Section 1 (Questions 1 – 19) requested the general demographics of the participants and their organisations. Respondents were able to input data or select from a variety of options for each question asked.
- Section 2 (Questions 20 – 78) asked the participant multiple questions in relation to research topic. The questions in this section were predominantly "Likert" scale structured questions where respondents were able to select one of five possible answers. Each answer, per question, was assigned a predetermine score from 1 to 5 that would then be utilised in the development of scales for analysis. The value for each answer is clearly described in chapter 4 of this research paper.

3.6 Method of Data Analysis

The data collected was analysed using descriptive and inferential statistics. Crosstabulation analysis and comparison of means was utilised to derived results across categories. Kruskal-Wallis tests were utilised to determine if there were statistically significant differences across categories and relationships were checked utilising Spearman's correlation analysis. The analysis was completed utilising both Microsoft Excel and IBM's statistical package for social sciences (SPSS) computer programme, version 25. The data generated was able to create a rich picture of the South African construction sector with regards to the topic of discussion.

3.7 Validity and Reliability of the Data

As mentioned previously all data sets received were analysed to ensure that the maximum and minimum values were within the defined limits and missing information was checked to ensure the validity of the data set was not compromised. Through this check of the data received 93 validated data sets were available for analysis.

When scales were developed that utilised data from multiple questions the reliability and internal consistency was examined utilising the Cronbach alpha coefficient analysis. Table 3 below shows the Cronbach alpha value developed for each scale developed:

Table 3: Reliability Statistics for all Scales Developed

	Cronbach's Alpha	N of Items
KM Maturity Level Indicator 1	.755	2
KM Maturity Level Indicator 2	.535	2
KM Maturity Level Indicator 3	.374	2
KM Maturity Level Indicator 4	.627	3
KM Maturity Level Indicator 5	.687	4
KM Maturity Level Indicator 6	.643	3
Key Barrier: Level of Training	.480	5
Key Barrier: Culture	.724	5
Key Barrier: Contractual Framework	.272	3
Key Barrier: Time	.339	3
Opinion Towards KM	.591	2
Opinion Towards Extending KM Activities Beyond Organizational Borders	.688	3

The internal consistency and reliability for most scales developed did not meet the .7 threshold to be considered acceptable (Pallant, 2011). However, as per Pallant (2011), as each scale has a relatively small number of items the mean inter-item correlation value would be a better tool to report reliability for these scales.

Therefore, for each scale the "Corrected Item-Total Correlation" value per question utilised in the scales make up was examined to ensure that none of the questions were measuring something other than intention of the scale overall. This is determined by ensuring that the "Corrected Item-Total Correlation" value is not below the threshold of .3 (Pallant, 2011). The result of this analysis is shown in Table 4 below:

Table 4: Item-Total Statistics for Each Scale Developed

Scale	Items	Corrected Item-Total Correlation	Cronbach's Alpha if Item Deleted
KM Maturity Level Indicator 1	Question 21	.610	.
	Question 26	.610	.
KM Maturity Level Indicator 2	Question 28	.369	.
	Question 33	.369	.
KM Maturity Level Indicator 3	Question 40	.232	.
	Question 48	.232	.
KM Maturity Level Indicator 4	Question 24	.468	.493

	Question 27	.535	.387
	Question 41	.332	.657
KM Maturity Level Indicator 5	Question 38	.600	.539
	Question 39	.069	.805
	Question 49	.622	.501
	Question 50	.626	.506
KM Maturity Level Indicator 6	Question 25	.343	.729
	Question 45	.633	.268
	Question 70	.437	.585
Key Barrier: Level of Training	Question 31	.038	.573
	Question 32	.252	.430
	Question 33	.561	.219
	Question 35	.402	.318
	Question 66	.118	.505
Key Barrier: Culture	Question 38	.516	.666
	Question 44	.381	.716
	Question 45	.225	.761
	Question 49	.664	.596
	Question 50	.647	.605
Key Barrier: Contractual Framework	Question 55	.096	.359
	Question 56	.177	.146
	Question 54	.193	.123
Key Barrier: Time	Question 30	.102	.426
	Question 42	.240	.151
	Question 57	.244	.149
Opinion Towards KM	Question 36	.426	.
	Question 65	.426	.
Opinion Towards Extending KM Activities Beyond Organizational Borders	Question 67	.602	.481
	Question 68	.353	.801
	Question 77	.586	.486

The analysis shows that the "Corrected Item-Total Correlation" values did fall below the .3 threshold for multiple scales (highlighted in red) and that by removing some questions from the scales the Cronbach alpha score could be improved per scale (highlighted in blue). However, as research in the social science is rarely precise and the topic of Knowledge Management is broad, it is possible that multiple questions on one topic could result in feedback that does not "hang together". An organisation may be strong in one aspect of KM and weak in another area, both which are considered a part of the same KM Maturity Level Indicator. The correlation of the feedback then to these questions can clearly then differ, resulting in the low scores developed. The decision was taken to keep the development of each scale utilising the questions listed, to utilise feedback from as many questions as possible.

3.8 Ethical Considerations

Before the survey was able to be distributed to any potential respondents the ethical considerations were examined. The survey would be distributed for Human subjects to complete and no harm was expected to be caused through their participation. No conflict of interested were expected and no form of discrimination was expected. The participants were given the option to consent to their feedback being utilised in this research and their confidentiality was honoured.

Upon review approval was given by the University of Cape Town faculty of Engineering and the Built Environment to proceed. A copy of the ethics approval can be found in Addendum B.

3.9 Summary of the Chapter

The Research Methodology chapter detailed the approach this research undertook to test the proposition posed. The postpositivist quantitative research approach was detailed and its use was justified. The population of the study was clarified and quantified utilising the latest data available. The methods for data analysis were then discussed with the validity and reliability of the data utilised for the research established. Finally the ethical consideration of the research were discussed with the agreement reach that this report would cause no harm to those that chose to participate.

4. Data Presentation, Analysis and Discussion

4.1 Introduction

In this chapter the data generated by the sample frame utilised for this research is presented, analysed and discussed. The data presented was generated by the respondent's feedback to the online survey circulated and represents their views on the questions posed regarding knowledge management and extending these activities beyond organisational borders.

4.2 Data Presentation

The primary data collected through the online survey is presented in the following subsections. It must be noted that feedback to every question was not compulsory and therefore there are cases where data is missing. The frequencies displayed are true to the data collected and the percentage listed is the valid percentage of data supplied, i.e. the missing data is excluded from the analysis where necessary.

4.2.1 Demographic Indicators of the Respondents

Based on the information received Table 5 was developed to show the overall demographic indicators of the respondents and their organisations. The feedback showed that the organisations represented in the survey predominantly provided Contracting, Consultancy and Project Management services. Of the Consultancy services provided, Mechanical and Electrical Engineering, were the most prevalent.

Participation in the survey was predominantly by employees of Micro and Small organisations with 55% of respondents being from organisations with 10 or fewer employees. Organisations were classed into categorical sizes using the latest SME definition in the construction sector by de Wet (2019) based on annual turnover.

Most organisations represented operated in the Western Cape (23%), Internationally (21%) and Nationally (17%). The respondents to the survey were predominantly between the ages of 35-44 (29%) and 45-54 (26%). The gender split was 77% Male, 13% Female, with 11% not specifying a gender. The breakdown per ethnicity showed most respondents to be either Black African (41%) or White (44%).

Table 5: Demographic Indicators of the Respondents and their Organisations

Demographic Indicator	Item	Frequency	Percentage (%)
Services Provided by Organization	Sub-Contractor	29	22
	Principal Contractor	24	18
	Project Management	23	17
	Consultant	22	17
	Building Resources Supplier	11	8
	Specialist	9	7
	Property Developer	7	5
	Quantity Surveyor	4	3
	Architect	3	2
Services Provided by Specialists & Consultants	Mechanical Engineering	4	24
	Electrical Engineering	3	18
	Civil Engineering	2	12
	Structural Engineering	2	12
	Health & Safety	2	12
	Fire Engineering	1	6
	Acoustic Engineering	1	6
	Academic	1	6
	Town Planning	1	6
Number of Employees in Partici-pants Organization	1 to 10 Employees	44	55
	11 to 20 Employees	10	13
	21 to 30 Employees	5	6
	31 to 40 Employees	2	3
	41 to 50 Employees	2	3
	51 to 60 Employees	1	2
	61 to 70 Employees	0	0
	71 to 80 Employees	1	2
	81 to 90 Employees	1	2
	91 to 100 Employees	2	3
	100+ Employees	12	15
Annual Turnover of Organization	R0mil to R10mil (Micro)	42	44
	R10mil to R75mil (Small)	14	15
	R75mil to 170mil (Medium)	8	8
	R170mil + (Large)	6	6
Region of Operation	Western Cape	22	23
	Internationally	20	21
	Nationally	16	17

	Gauteng	10	11
	North West	4	4
	Eastern Cape	3	3
	Mpumalanga	2	2
	Limpopo	2	2
	Not Specified	2	2
	Free State	1	1
	Northern Cape	1	1
Age of Respondents	18 – 24	2	2
	24 – 34	16	17
	35 – 44	27	29
	45 – 54	25	26
	55 – 64	16	17
	65 +	8	8
	Not Specified	1	1
Gender of Respondents	Female	12	13
	Male	73	77
	Not Specified	10	11
Ethnicity of Respondents	White	42	44
	Black African	39	41
	Not Specified	9	10
	Coloured	3	3
	Asian	1	1
	Indian	1	1

4.2.2 KM Maturity Level Indicator of Participants Organisation

The first objective of this research was to establish an indicator of what the KM maturity level of each participants organisation was based on their feedback on specific questions. The KM maturity model that was utilised for this research was from Kruger and Snyman (2007) as shown in Table 2.

It was necessary to have an indication of the KM maturity level of each organisation for the analysis of the key barriers identified in the later sections of this research. It was beyond the scope of this research to complete a thorough analysis of each organisations processes and activities in order to determine their exact level of KM maturity. However, based on the feedback received from the surveyed participants, an indication could be derived for each organisation represented.

As the focus of this paper was to evaluate if an organisation extended their KM activities beyond organisational borders the Author determined that it was not necessary to determine if their maturity had progressed to Kruger and Snyman (2007) 7th and final KM maturity level. For the data collection and analysis of this paper the 7th level of KM maturity was combined into the 6th level. It was not expected that many organisations would have achieved the 7th level of KM maturity and therefore to reflect this as a separate level would potentially dilute the results unnecessarily.

A selection of questions was picked for each KM maturity level and a combined score was determined where more than one question was utilised. This combined score was determined by a summed average for all questioned utilised. A "gate" style analysis was utilised to then qualify for each KM maturity level. To qualify for a level the organisation had to score 4 or above for that level and also had to have qualified for all preceding KM maturity levels. A score of 4 was selected as the qualifying criteria as that would indicate absolute agreement for all questions posed related to that level.

The development of the KM Maturity Level Indicator for each organisation is shown in Table 6 and the summary of the assigned indicators to each organisation is shown in Table 7.

Table 6: KM Maturity Level Indicator Results

	Questions Utilized	Key Factor	Frequency Bins					Average
			1	2	3	4	5	
Level 1	Question 21 (Y=5 N=1)	Tacit knowledge captured and stored electronically	37				56	3.4
	Question 26 (Y=5 N=1)	Explicit knowledge captured and stored electronically	24				65	3.9
		Combined Score	24		17		54	3.7
41 Organisations therefore failed to qualify for a KM Maturity Level Indicator of 1								
54 Organisations qualified for a KM Maturity Level Indicator of Level 1								
Level 2	Question 28 (VL=5 VU=1)	Explicit knowledge is available to all employees.	3	9	18	34	26	3.8
	Question 33 (SA=5 SD=1)	KM implemented through all levels of organization.	2	6	16	40	14	3.7
		Combined Score*				23	12	
35 Organisations qualified for a KM Maturity Level Indicator of Level 2								
Level 3	Question 40 (SA=5 SD=1)	Knowledge is a strategic resource	1	3	7	29	39	4.3
	Question 48 (SA=5 SD=1)	KM is driven by upper management	3	6	12	39	19	3.8
		Combined Score*				16	9	
25 Organisations qualified for a KM Maturity Level Indicator of Level 3								
Level 4	Question 24 (VL=5 VU=1)	Formalized coaching/mentoring	11	16	11	35	22	3.4
	Question 27 (VL=5 VU=1)	Training documentation created from KM	3	17	12	34	24	3.7

	Question 41 (SA=5 SD=1)	KM has developed new strategies/processes	3	3	11	42	20	3.9
		Combined Score*				13	7	
	**20** Organisations qualified for a KM Maturity Level Indicator of Level 4							
Level 5	Question 38 (VU=5 VL=1)	Culture of not discussing mistakes	7	14	15	31	10	3.3
	Question 39 (SA=5 SD=1)	KM created a culture of sharing knowledge	1	8	14	40	16	3.8
	Question 49 (VU=5 VL=1)	Internal competition limits KM	10	16	12	23	18	3.3
	Question 50 (VU=5 VL=1)	Job security limits KM	14	11	14	25	13	3.2
		Combined Score*				8	3	
	**11** Organisations qualified for a KM Maturity Level Indicator of Level 5							
Level 6	Question 25 (VL=5 VU=1)	Lessons are shared beyond borders	10	20	17	33	15	3.2
	Question 45 (VL=5 VU=1)	Encouraged to interact with peers	3	7	6	32	31	4.0
	Question 70 (VL=5 VU=1)	Benefitted from relationships	1	3	5	36	21	4.1
		Combined Score*				7	3	
	**10** Organisations qualified for a KM Maturity Level Indicator of Level 6							

* The combined score shows only the data for respondents that qualified for that KM Maturity Level Indicator by scoring 4 or above and having qualified for the previous KM Maturity Level Indicator. Values were rounded down into the closest frequency bin to not clutter the display of information.

Table 7: Summary of Assigned KM Maturity Level Indicators

KM Maturity Level Indicator	Frequency	Percent (%)
0	41	43
1	19	20
2	10	11
3	5	5
4	9	10
5	1	1
6	10	11

4.2.3 Key Barriers

The next objective of this research was to determine the perception the respondents has towards the impact the identified key barriers had on KM activities and extending these beyond organisational borders.

4.2.3.1 Key Barrier 1: The Level of Knowledge/Training in KM Provided by Construction Organisations

The first key barrier to be examined in this research was to evaluate the level of KM training given by each organisation to its employees through feedback received on certain questions. A score was developed for each respondent through a summed average for the questions listed in Table 8. The data collected in respect of this barrier is presented in Table 8 below:

Table 8: Perception of Knowledge/Training Received

Questions Utilized	Key Factor	Frequency Bins					Average
		1	2	3	4	5	
Question 31 (SD=5 SA=1)	KM is implemented Ad Hoc	6	35	18	13	7	2.8
Question 32 (VL=5 VU=1)	KM is implemented throughout project	5	8	16	34	15	3.6
Question 33 (SA=5 SD=1)	KM is implemented through all organizational levels	2	6	16	40	14	3.7
Question 35 (SD=5 SA=1)	KM is only for upper management	5	20	13	30	11	3.3
Question 66 (SD=5 SA=1)	All projects are different	1	5	8	25	27	4.1
Combined Score*		0	13	50	16	0	3.5
Cumulative Percentage (%)		0	17	80	100	0	

*** The combined score shows the scores collapsed down into the closest frequency bin to not clutter the display of information.**

4.2.3.2 Key Barrier 2: Establish Whether the Culture Within a Construction Organisation is Conducive to Sharing Knowledge

The second key barrier to be examined was to evaluate how conducive the culture was in each participants organisation to sharing knowledge. A score was developed for each respondent through a summed average for the questions listed in Table 9. The data collected in respect of this barrier is presented in Table 9 below:

Table 9: Perception of How Conducive the Organisational Culture is for KM

Questions Utilized	Key Factor	Frequency Bins					Average
		1	2	3	4	5	
Question 38 (VU=5 VL=1)	Mistakes are not recorded	7	14	15	31	10	3.3
Question 44 (VU=5 VL=1)	Restructuring initiatives limits KM	6	22	20	18	13	3.1
Question 45 (VL=5 VU=1)	Encouraged to engage with peers	3	7	6	32	31	4.0
Question 49 (VL=5 VU=1)	Internal competition limits KM	10	16	12	23	18	3.3
Question 50 (VU=5 VL=1)	Job security limits KM	14	11	14	25	13	3.2
	Combined Score*	1	18	37	21	2	3.4
	Cumulative Percentage (%)	1	24	71	98	2	

*** The combined score shows the scores collapsed down into the closest frequency bin to not clutter the display of information.**

4.2.3.3   Key Barrier 3: Determine Whether the Contractual Framework of Construction Projects was Conducive to Knowledge Sharing

The third key barrier to be examined in this research was to evaluate how conducive the contracts utilised in the construction sector were to enable knowledge sharing amongst all stakeholders involved. A score was developed for each respondent through a summed average for the questions listed in Table 10. The data collected in respect of this barrier is presented in Table 10 below:

Table 10: Perception of How Conducive Contracts are to KM

Questions Utilized	Key Factor	Frequency Bins					Average
		1	2	3	4	5	
Question 54 (SD=5 SA=1)	Contracts limit communication	2	19	26	21	4	3.1
Question 55 (VL=5 VU=1)	Project sponsor encourages KM	9	12	12	26	12	3.3
Question 56 (VU=5 VL=1)	Contracts limit knowledge transfer	2	23	17	27	3	3.1
	Combined Score*	2	22	37	10	1	3.1
	Cumulative Percentage (%)	3	33	85	99	100	

*** The combined score shows the scores collapsed down into the closest frequency bin to not clutter the display of information.**

The fourth key barrier to be examined in this research was to determine if construction organisations are incentivising their staff to interact and share knowledge, in addition to their normal project activities. This barrier utilised feedback from only one question and the data collected in respect of this barrier is presented in Table 11 below:

Table 11: Perception Towards KM Incentivisation

Questions Utilized	Key Factor	Frequency Bins					Average
		1	2	3	4	5	
Question 46 (VL=5 VU=1)	Organisation is incentivizing employees	7	11	18	21	20	3.5
	Cumulative Percentage (%)	9	23	47	74	100	

4.2.3.5   Key Barrier 5: Establish if Enough Time was Allocated for KM Activities within Construction Projects.

The fifth key barrier to be examined in this research was to evaluate if enough time is given to employees to complete KM activities during their construction projects. A score was developed for each respondent through a summed average for the questions listed in Table 12. The data collected in respect of this barrier is presented in Table 12 below:

Table 12: Perception of if Enough Time was Allocated to KM

Questions Utilized	Key Factor	Frequency Bins					Average
		1	2	3	4	5	
Question 30 (SD=5 SA=1)	KM activities completed in spare time	3	21	20	25	9	3.2
Question 42 (SD=5 SA=1)	Knowledge is regularly lost when not captured in time	10	32	18	15	4	2.6
Question 57 (VU=5 VL=1)	Project time pressure limits time for KM activities	10	31	17	11	3	2.5
	Combined Score*	4	38	31	6	0	2.8
	Cumulative Percentage (%)	5	53	92	100		

*** The combined score shows the scores collapsed down into the closest frequency bin to not clutter the display of information.**

4.2.3.6   Establish the Factors Behind the Key Barriers in Order of Prevalence and Severity of their Impact on Knowledge Sharing and Growing KM Maturity within Construction Organisations.

The third objective of this research was to determine the prevalence and severity of factors that would have an influence on the key barriers identified. The factors selected for analysis were as follows; Time,

KM Training, Management Support, Continuity, Contractual Framework, Job Security, Language Challenges, Temporary Multi-Organisation Coalition Challenges.

For each factor a combined mean score was developed through a summed average for the questions utilised. The development of the score for each factor is shown in Table 13 below and the ranked results are shown in Table 14 below:

Table 13: Development of Factor Score

Factor	Questions Utilized	Key Factor	Frequency Bins					Average
			1	2	3	4	5	
KM Training	Question 34 (SA=5 SD=1)	Insufficient training limits KM	7	23	17	28	4	3.0
		Cumulative Percentage (%)	9	38	60	95	100	
Time	Question 30 (SA=5 SD=1)	KM activities in spare time	9	25	20	21	3	2.8
	Question 42 (SA=5 SD=1)	Knowledge regularly lost	4	15	18	32	10	3.4
	Question 57 (VL=5 VU=1)	Not enough time allocated for KM	3	11	17	31	10	3.5
		Combined Score*	1	20	41	15	2	3.3
		Cumulative Percentage (%)	1	27	79	98	100	
Job Security	Question 38 (VL=5 VU=1)	Mistakes not wanted to be recorded	10	31	15	14	7	2.7
	Question 44 (VL=5 VU=1)	Business restructuring limits knowledge flow	13	18	20	22	6	2.9
	Question 50 (VL=5 VU=1)	Job security limits knowledge flow	13	25	14	11	14	2.8
	Question 62 (VL=5 VU=1)	Antagonism between age groups limits KM	8	35	13	15	1	2.5
		Combined Score*	10	33	24	11	1	2.8
		Cumulative Percentage (%)	13	54	85	99	1	
TMO Challenges	Question 58 (VU=5 VL=1)	Difficult to arrange for the full project team to be in one meeting together	6	43	15	6	2	2.4
		Cumulative Percentage (%)	8	68	89	97	100	

Continuity	Question 37 (SA=5 SD=1)	KM limited by high staff turn-over	14	31	16	14	4	2.5
	Question 43 (SA=5 SD=1)	KM Limited as staff move onto other projects before KM activities	6	23	20	26	4	3.0
	Question 59 (VL=5 VU=1)	Project team continuity rare from project to project	4	17	22	19	10	3.2
		Combined Score*	5	37	24	12	1	2.9
		Cumulative Percentage (%)	6	53	84	99	100	
Management Support	Question 47 (SA=5 SD=1)	More resources needed for KM activities	3	5	18	34	19	3.8
	Question 48 (SD=5 SA=1)	Management drive KM	19	39	12	6	3	2.2
		Combined Score*	2	15	56	5	1	3.0
		Cumulative Percentage (%)	3	22	92	99	1	
Language	Question 51 (VL=5 VU=1)	Multiple language spoken in organization limit knowledge transfer	20	27	10	13	8	2.5
	Question 61 (VL=5 VU=1)	Multiple languages spoken in projects limit knowledge transfer	14	30	15	9	4	2.4
		Combined Score*	20	28	15	10	6	2.6
		Cumulative Percentage (%)	25	61	80	92	100	
Contractual Framework	Question 54 (SA=5 SD=1)	Contracts limit communication	4	21	26	19	2	2.9
	Question 55 (VU=5 VL=1)	Project sponsor encourages collaboration	12	26	12	12	9	2.7
	Question 56 (VL=5 VU=1)	Contracts limit knowledge transfer	3	27	17	23	2	2.9
		Combined Score*	5	30	32	4	1	2.9
		Cumulative Percentage (%)	7	49	93	99	100	

*** The combined score shows the scores collapsed down into the closest frequency bin to not clutter the display of information.**

Table 14: Ranked List of Factors

Factor	Ranking	Mean Score
Time	1	3.3
KM Training	2	3.0
Management Support	3	3.0
Continuity	4	2.9
Contractual Framework	5	2.9
Job Security	6	2.8
Language	7	2.6
TMO Challenges	8	2.4

4.2.4 Current Opinion Towards KM and Extending KM Activities Beyond Organisational Borders

The fourth and final objective of this research was to determine an indication of what the current opinion is towards Knowledge Management and extending KM activities beyond organisational borders by those working in the South African construction sector. This objective was researched by first determining a scale for the respondent's opinion towards Knowledge Management as a strategy, followed by developing a scale for their opinion on extending KM activities beyond organisational borders.

4.2.4.1 The Current Opinion Towards KM in the South African Construction Sector

A combined mean score was developed through a summed average for questions utilised. The data collected in respect of this scale is presented in Table 15 below:

Table 15: Opinion Towards KM

Questions Utilized	Key Factor	Frequency Bins					Average
		1	2	3	4	5	
Question 36 (SD=5 SA=1)	KM is a managerial fad.	2	11	18	36	12	3.6
Question 65 (SA=5 SD=1)	Effective KM is valuable.	0	2	4	27	31	4.4
	Combined Score*	2	8	15	46	8	3.8
	Cumulative Percentage (%)	3	13	32	90	100	

*** The combined score shows the scores collapsed down into the closest frequency bin to not clutter the display of information.**

A combined mean score was developed through a summed average for questions utilised. The data collected in respect of this scale is presented in Table 16 below:

Table 16: Opinion Towards Extending KM Activities Beyond Organisational Borders

Questions Utilized	Key Factor	Frequency Bins					Average
		1	2	3	4	5	
Question 67 (VL=5 VU=1)	Organisation would benefit from external knowledge	2	1	8	31	24	4.1
Question 68 (SD=5 SA=1)	Unlikely to learn from peers	3	4	8	25	26	4.0
Question 73 (VL=5 VU=1)	Relationships can uplift the whole construction sector	3	1	7	31	24	4.1
	Combined Score*	1	2	12	37	11	4.1
	Cumulative Percentage (%)	2	5	27	83	100	

*** The combined score shows the scores collapsed down into the closest frequency bin to not clutter the display of information.**

4.2.5 Additional Feedback

Included into the survey were a few questions related to the topic of extending KM activities beyond organisational borders that were not used in the development of the previous objectives but provides further insight into the topic. These questions were included in order to gain insight into broader areas of the overall topic as follows. The data collected in respect of these questions is presented in Table 17 below:

Table 17: Additional Questions

Questions Utilized	Key Factor	Frequency Bins					Average
		1	2	3	4	5	
Question 57 (VU=5 VL=1)	Time pressure limits KM in projects	10	31	17	11	3	3.5
	Cumulative Percentage (%)	14	57	81	96	100	
Question 64 (VU=5 VL=1)	Compromised competition	6	21	13	13	12	3.1
	Cumulative Percentage (%)	9	42	62	82	100	
Question 71 (VL=5 VU=1)	Project team continuity	2	2	9	33	20	4.0
	Cumulative Percentage (%)	3	6	20	70	100	
Question 72 (SD=5 SA=1)	Learning on the job	9	27	13	16	1	2.6

		Cumulative Percentage (%)	14	55	74	99	100	
Question 76 (SA=1 SD=1)	Knowledge leaving SA		4	0	14	22	25	4.0
		Cumulative Percentage (%)	6	6	28	62	100	
Question 77 (SD=5 SA=1)	Disciplinary action		7	10	20	13	16	3.3
		Cumulative Percentage (%)	11	26	56	76	100	

The final additional question posed to respondents was as follows:

In your opinion which of the following prevalent construction project issues could be addressed by organisations working closer together and transferring knowledge freely:

- *Coordination of information*
- *Conflict resolution*
- *Planning & scheduling*
- *Cash-flow management*
- *Resource availability*
- *Complexities of new technologies*
- *Abortive works*

The feedback received was tabulated in Table 18 as follows:

Table 18: Feedback received for which construction project issues could be Addressed through better collaboration

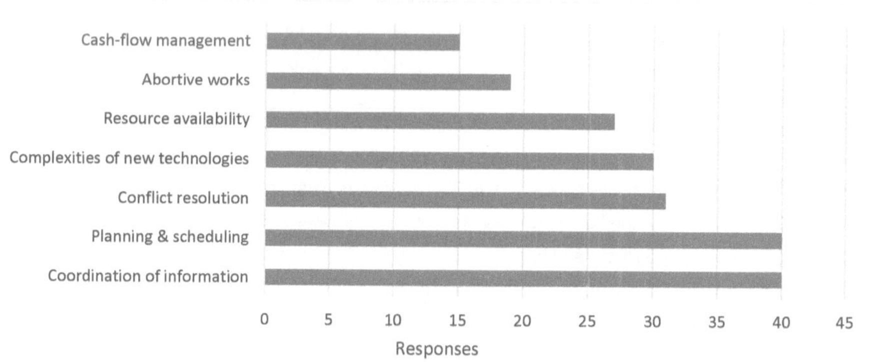

4.3 Data Analysis

The following section for this research describes the analysis completed on the data in order to generate findings from the data collected.

4.3.1 KM Maturity Level Indicator

Whilst research by Kruger and Johnson (2009) had found that the growth of KM maturity in South African construction organisations was most significant in larger organisations a Kruskal-Wallis Test on the resultant KM Maturity Level Indicator developed by this research failed corroborate their findings. The test revealed no statistically significant difference in the turnover of organisations across the KM Maturity Levels x^2 (5, n = 70) = 1.21, p = .94 (Pallant, 2011).

4.3.2 Key Barriers

Based on the data presented the following analysis was carried out on the key barriers identified.

4.3.2.1   Key Barrier 1: The Level of Knowledge/Training in KM provided by Construction Organisations

By observing the results shown in Table 8, with an overall mean score of 3.5 and 83% of respondents having a score of 3 or greater, indicates that most respondents believe their organisations are providing enough knowledge and training to their employee's on the topic of Knowledge Management.

The largely positive feedback indicating that most respondents believed they received enough training in KM was verified by testing its correlation to Question 34; where the participant was asked:

> *KM in my organisation is limited in its implementation as employees do not have sufficient training/understanding/expertise to carry out activities effectively? (SD= 5 SA=1)*

Using Spearman's correlation analysis there was a medium correlation found between the two variables, *rho* = .446 as shown in Table 23 (Pallant, 2011).

A crosstabulation analysis (Table 25 found in Addendum C), Kruskal-Wallis test and comparison of means between this barrier and the KM Maturity Level Indicator was completed (as shown in Table 22 and Table 24 at the end of this section). The Kruskal-Wallis Test revealed a statistically significant difference in the scores for the barrier across the KM Maturity Levels, x^2 (6, n = 79) = 31.018, p = .001 (Pallant, 2011). A small positive trend found is evident (M=0.22) when the mean score per KM Maturity Level is examined with the more mature organisations scoring higher. This result is encouraging and lends credibility to the notion that the amount of knowledge and training a respondent had on KM would increase with the KM maturity level of their organisation. It must be noted that the organisations with lower KM maturity levels scored in the region of 3.2 indicating near neutrality in regard to the training received on KM. When examining the crosstabulation analysis (as shown in Table 25 found in Addendum C) it was found that 16.5% of respondents scored below 3 and therefore indicated that they did not receive sufficient training in KM. The breakdown of all these respondents per KM Maturity Level was as follows:

- 35.5% of respondents from an organisation with KM Maturity Level 0 scored below 3.
- 13.4% of respondents from an organisation with KM Maturity Level 1 scored below 3.

Based on the overall score and the trend found this barrier was less prevalent in mature organisations and therefore it is less likely to be prevalent in inhibiting KM activities from extending beyond organizational borders in mature organisations. This is due to the definition of the maturity levels shown in Table 2 where extending KM activities beyond organizational borders is a trait of the 6th KM maturity level.

4.3.2.2 Key Barrier 2: Establish Whether the Culture Within a Construction Organisation is Conducive to Sharing Knowledge

By observing the results shown in Table 9, with an overall mean score of 3.4 and 75% of participants having a score of 3 or greater, indicates that most organisations have created a culture that is conducive to sharing knowledge internally.

To establish if the implementation of KM had an impact on this positive culture of sharing knowledge result a correlation analysis was completed. The resultant scale developed for the barrier was compared to Question 39; where the participant was asked:

> *KM in my organisation is creating a new culture of sharing knowledge in your organization. (SA=5 SD=1)*

Interestingly when using Spearman's correlation analysis as shown in Table 23 there was only a small positive correlation found between the two variables, $rho = .238$, $n = 79$, $p = .04$ (Pallant, 2011).

A crosstabulation analysis (as shown in Table 26 in Addendum C), Kruskal-Wallis test and comparison of means between this barrier and the KM Maturity Level Indicator was completed (as shown in Table 22 and Table 24). The Kruskal-Wallis Test revealed a statistically significant difference in scores for the barrier across the KM Maturity Levels, $x^2 (6, n = 79) = 24.47$, $p = .001$ (Pallant, 2011). A positive trend found was evident (M= 0.22) when the mean score per KM Maturity Level is examined with the more mature organisations scoring higher. This result confirms the thought that the culture of an organisation to share knowledge improves with their KM maturity level. Upon examination of the scores in the crosstabulation analysis in Table 26 it was found that 24.1% of respondents scored below 3 and therefore indicated that the culture within their organisation was not suitable to sharing knowledge. The breakdown per KM maturity level was as follows:

- 35.5% of respondents from an organisation with KM Maturity Level 0 scored below 3.
- 33.3% of respondents from an organisation with KM Maturity Level 1 scored below 3.
- 12.5% of respondents from an organisation with KM Maturity Level 2 scored below 3.
- 22.2% of respondents from an organisation with KM Maturity Level 4 scored below 3.

Based on the score and trend found this barrier was therefore less likely to be prevalent in inhibiting KM activities from extending across organisational borders in the mature organisations and only had a modest impact on the less mature organisations.

4.3.2.3   Key Barrier 3: Determine Whether the Contractual Framework of Construction Projects was Conducive to Knowledge Sharing

As per Table 10 the result of this analysis showed an overall mean score of 3.1 indicating a positive response to the question posed. Interestingly the breakdown of the results showed that 33% of participants scored below 3 and therefore had a negative perception towards the contractual framework fostering an environment towards knowledge sharing. Only 15% of participants scored higher than 4 and therefore had a clear positive perception towards the objective. The remaining participants scored between 3 and 4 – displaying a largely neutral perception towards the objective.

A correlation analysis was conducted between this barrier and Question 60; where the participants were asked:

> *In the typical construction project, the threat of penalties for causing project delays limits collaboration between virtual team members. (VU=5 VL=1)*

It would have been thought that respondents that didn't believe the current contract structure is conducive to sharing knowledge would also believe that the threat of penalties in these contracts would be contributing to this lack of collaboration. However, as shown in Table 23 the result of a Spearman correlation analysis showed only a small, positive correlation between the two variables, *rho* = .10, *n* = 72, *p* = .39 (Pallant, 2011).

A crosstabulation analysis and comparison of means between this barrier and the KM maturity level was not conducted as it was thought by the Author that the contracts entered into by construction organisations would not be dependent on their KM maturity level.

4.3.2.4   Key Barrier 4: To Determine if Construction Organisations were Incentivizing Resources to Interact and Share Knowledge, in Addition to their Normal Project Activities.

By observing the results shown in Table 11 an overall mean score of 3.5 indicates an overall positive response to the question posed. When considering the breakdown of the results it was found that 23% of participants indicated that employees were not likely, or very unlikely, to be incentivised to interact and share knowledge beyond their normal project duties, 24% of participants were neutral and the remaining 53% of participants confirmed this was likely or very likely to occur.

A crosstabulation analysis (as shown in Table 27 in Addendum C), Kruskal-Wallis Test and comparison of means was conducted between this barrier and the KM Maturity Level Indicator was completed (as shown in Table 22 and Table 24). The Kruskal-Wallis Test revealed a statistically significant difference in the scores for the barrier across the KM Maturity Levels, x^2 (6, *n* = 77) = 16.01, *p* = .01 (Pallant, 2011). A small positive trend found is evident (M= 0.25) when the mean score per KM Maturity Level in Figure 1 is examined with the more mature organisations scoring higher. When examining the details of the crosstabulation analysis in Table 27 it was found that 23.4% of respondents scored below 3 and therefore indicated that they were not incentivised to share knowledge beyond their normal work activities. The breakdown of these respondents per KM Maturity Level is as follows:

- 36.7% of respondents from an organisation with KM Maturity Level 0 scored below 3.
- 26.7% of respondents from an organisation with KM Maturity Level 1 scored below 3.
- 50.0% of respondents from an organisation with KM Maturity Level 3 scored below 3.
- 10.0% of respondents from an organisation with KM Maturity Level 6 scored below 3.

Based on the overall score and trend found this barrier was therefore less likely to inhibit KM activities internally and from extending across organisational borders in mature organisations.

4.3.2.5 Key Barrier 5: Establish if Enough Time is Allocated for KM Activities Within Construction Projects.

As per Table 12 the result of this analysis showed an overall mean score of 2.8 indicating a negative outlook to the barrier. The breakdown of the results showed that 53% of participants recorded a score of below 3 indicating that based on the questions posed they did not believe enough time is allocated to KM activities.

A crosstabulation analysis (as shown in Table 28 in Addendum C), Kruskal-Wallis test and comparison of means between Objective 5 and the KM Maturity Level Indicator was completed (as shown in Table 22 and Table 24). The Kruskal-Wallis Test revealed no statistically significant difference in the scores for the barrier across the KM Maturity Levels, x^2 (6, n = 79) = 12.05, p = .061 (Pallant, 2011). The lack of a statistically significant relationship is shown when examining the mean score per KM Maturity Level in Figure 1 with only a very small positive trend found (M=0.10). When examining the details of the crosstabulation analysis in Table 28 it was found that 53.2% of respondents scored below 3 and therefore indicated that not enough time was allocated for KM activities within their projects. The breakdown of all these respondents, per KM Maturity Level, is as follows:

- 54.8% of respondents from an organisation with KM Maturity Level 0 scored below 3.
- 66.6% of respondents from an organisation with KM Maturity Level 1 scored below 3.
- 37.5% of respondents from an organisation with KM Maturity Level 2 scored below 3.
- 80.0% of respondents from an organisation with KM Maturity Level 3 scored below 3.
- 55.5% of respondents from an organisation with KM Maturity Level 4 scored below 3
- 30.0% of respondents from an organisation with KM Maturity Level 6 scored below 3

The negligent positive trend and low overall score indicates that the barrier is an issue regardless of KM maturity and that this barrier will inhibit KM activities from extending across organisational borders even in the more mature organisations.

4.3.3   Factors that Influence the Key Barriers

Based on the data presented the following analysis was carried out on the factors identified.

4.3.3.1   Time Factor

With 73% of respondents scoring 3 or higher for this factor and an overall mean score of 3.3 the time allocated to KM activities was identified as the most significant factor researched. With a mean score of over 3 this factor is a clear barrier to KM's implementation. Further, as discovered in the development of the Time Key Barrier earlier in this research, it was found that there was no clear trend to suggest if more mature or less mature organisations provide enough time for KM activities. By examining the mean score for this barrier per KM Maturity Level Indicator in Table 24 all levels except levels 2 and 5 scored above 3. However, as level 5 only has 1 organisation in its category this result can be discounted. The barrier of providing enough time to complete KM activities can therefore be argued to be an across the board issue and will be a factor inhibiting KM activities from extending across organisational borders. As the same questions were utilised in the development of the Time Key Barrier as the Time Factor a second crosstabulation analysis was not needed – the analysis completed for the Time Key Barrier for correlation and statistical significance could be pulled through.

4.3.3.2   KM Training Factor

Feedback to whether the amount of training given on KM was a barrier to its implementation resulted in a mean score of 3.0 and was ranked 2nd highest of the factors researched. The score of 3.0 indicates that overall the respondents were on the fence as to whether this is a clear barrier to KM's implementation and extending KM activities beyond organisational borders.

A Kruskal-Wallis test and comparison of means between this factor and the KM Maturity Level Indicator was completed (as shown in Table 22 and Table 24). The Kruskal-Wallis Test revealed a statistically significant difference in the KM Training factor across the KM Maturity Level Indicator, x^2 (6, n = 79) = 18.46, p = .01 (Pallant, 2011). A very slight negative trend (M=-0.06) was evident when the mean score per KM Maturity Level Indicator was examined with the less mature organisations scoring higher. Interestingly upon examination of the crosstabulation analysis (Table 29 in Addendum C) 40.5% of respondents indicated that KM Training was a factor by scoring above 3, this is a significant proportion of respondents and shouldn't be ignored. The breakdown of all these respondents, per KM Maturity Level, is as follows:

- 67.8% of respondents from an organisation with KM Maturity Level 0.
- 20.0% of respondents from an organisation with KM Maturity Level 1.
- 25.0% of respondents from an organisation with KM Maturity Level 2.
- 20.0% of respondents from an organisation with KM Maturity Level 3.
- 33.3% of respondents from an organisation with KM Maturity Level 4.
- 20.0% of respondents from an organisation with KM Maturity Level 6.

This feedback ties back into analysis completed for the Knowledge/Training Key Barrier discussed earlier in this research where it was determined that the more mature organisations provided more training to their staff on KM and therefore this factor will be less likely to inhibit KM activities bother internally and from extending across organisational borders in more mature organisations. .

4.3.3.3 Management Support Factor

Feedback as to whether management provided adequate support to KM's implementation resulted in a mean score of 3.0 ranking 3rd in the list of factors researched. The score of 3 indicates that the respondents were neutral when indicating if management support inhibited KM's implementation overall. An examination of the breakdown of the feedback shows this neutrality and is clearly highlighted with 70% of respondents feedback score falling into the "3" frequency bin.

A Kruskal-Wallis test and comparison of means between this factor and the KM Maturity Level Indicator was completed (as shown in Table 22 and Table 24). The Kruskal-Wallis Test revealed a statistically significant difference in the Management Support barrier across the KM Maturity Levels, x^2 (6, n = 79) = 15.04, p = .02 (Pallant, 2011). A negative trend was found (M=-0.13) when the mean score per KM Maturity Level was examined with the less mature organisations scoring higher. Through a crosstabulation analysis (Table 33 in Addendum C) an examination showed of the 21.5% of respondents that scored below 3 and therefore indicated that Management Support was a barrier per KM Maturity Level was as follows:

- 32.4% of respondents from an organisation with KM Maturity Level 0.
- 6.7% of respondents from an organisation with KM Maturity Level 1.
- 50.0% of respondents from an organisation with KM Maturity Level 2.
- 33.3% of respondents from an organisation with KM Maturity Level 4.
- 10.0% of respondents from an organisation with KM Maturity Level 6.

Therefore, this factor will less likely inhibit KM activities internally and from extending across organisational borders in more mature organisations, with the only significant impact being shown in the less mature organisations.

4.3.3.4 Continuity Factor

Feedback as to whether the continuity of project teams from project to project was a factor to limit KM activities resulted in a mean score of 2.9 and ranked 4th in the list of factors researched. This score was just below the neutral score of 3 and indicates neutrality towards this factor by most respondents.

A Kruskal-Wallis test and comparison of means between this factor and the KM Maturity Level Indicator was completed (As shown in Table 22 and Table 24). The Kruskal-Wallis Test revealed no statistically significant difference in the Continuity factor across the KM Maturity Levels, x^2 (6, n = 79) = 9.73, p = .14 (Pallant, 2011). However, when examining the means per KM Maturity Level Indicator a very slight negative trend (M=-0.09) was found between the factor and KM maturity levels. A crosstabulation analysis between this factor and the KM Maturity Level Indicators (as shown in Table 32 in Addendum C) showed that the breakdown of the 36.7% respondents who thought this was a barrier to be as follows:

- 45.2% of respondents from an organisation with KM Maturity Level 0.
- 46.7% of respondents from an organisation with KM Maturity Level 1.
- 25.0% of respondents from an organisation with KM Maturity Level 2.
- 40.0% of respondents from an organisation with KM Maturity Level 3.
- 33.3% of respondents from an organisation with KM Maturity Level 4.
- 10.0% of respondents from an organisation with KM Maturity Level 6.

Interestingly significant percentages of respondents indicated this factor was a concern in organisations in both the low and middle KM maturities. However, the lack of an appreciable trend and overall neutral score indicates that this is not a significant factor inhibiting KM activities or stopping them from extending across organisational borders in the more mature organisations.

4.3.3.5 Contractual Framework Factor

Ranking 5th out of the factors researched, with a mean score of 2.9 was the impact that the contracts had on knowledge transfer. This factor scored close to the neutral ground of 3 and indicates that most respondents were neutral as to whether the contracts utilised in construction projects inhibited KM activities.

A Kruskal-Wallis test and comparison of means between this factor and the KM Maturity Level Indicator was completed (as shown in Table 22 and Table 24). The Kruskal-Wallis Test revealed no statistically significant difference in the Contractual framework barrier across the KM Maturity Levels, x^2 (6, n = 72) = 4.10, p = .66 (Pallant, 2011). The lack of a statistically significant difference across the maturity levels can be expected as per the earlier analysis for the Contractual Framework Key Barrier. It was stated earlier that the contracts utilised by construction organisation would not differ due to KM maturity and therefore this should not have an impact on the respondent's opinion towards the contracts and their influence on KM activities. The examination of the mean score per KM Maturity Level shows a uniform set of results with only a negligible negative trend found (M=-0.01). Based on a crosstabulation analysis between this factor and the KM Maturity Level Indicator (as shown in Table 35 in Addendum C) the breakdown of the 33.3% of respondents who considered this factor to be a barrier was as follows:

- 37.0% of respondents from an organisation with KM Maturity Level 0.
- 40.0% of respondents from an organisation with KM Maturity Level 1.
- 14.3% of respondents from an organisation with KM Maturity Level 2.
- 25.0% of respondents from an organisation with KM Maturity Level 3.
- 25.0% of respondents from an organisation with KM Maturity Level 4.
- 40.0% of respondents from an organisation with KM Maturity Level 6.

The lack of an appreciable trend and neutral overall score indicates that this is not a significant factor inhibiting KM activities or stopping them from extending across organisational borders for all KM maturities.

4.3.3.6   Job Security Factor

The 6th ranked factor was Job Security with a mean score of 2.8 based on the feedback received from the respondents. With 67% of respondents scoring below 3 most respondents did not believe that KM activities were being inhibited by the respondents need to protect their jobs.

A Kruskal-Wallis test and comparison of means between this factor and the KM Maturity Level Indicator was completed (as shown in Table 22 and Table 24). The Kruskal-Wallis Test revealed a statistically significant difference in the Job Security factor across the KM Maturity Levels, x^2 (6, n = 79) = 15.13, p = .02 (Pallant, 2011). A negative trend was found (M=-0.22) when the mean score per KM Maturity Level Indicator was examined with the less mature organisations scoring higher. Based on a crosstabulation analysis between this factor and the KM Maturity Level Indicators (as shown in Table 30 in Addendum C) the breakdown of the 29.1% of respondents who considered this to be a barrier was as follows:

- 41.9% of respondents from an organisation with KM Maturity Level 0.
- 26.7% of respondents from an organisation with KM Maturity Level 1.
- 37.5% of respondents from an organisation with KM Maturity Level 2.
- 20.0% of respondents from an organisation with KM Maturity Level 3.
- 22.2% of respondents from an organisation with KM Maturity Level 4.

Therefore, this factor only shows an impact in the less mature organisations and will less likely inhibit KM activities internally and from extending across organisational borders in more mature organisations.

4.3.3.7   Language Factor

The 7th ranked factor was Language with a mean score of 2.6 based on the feedback received from the respondents. With 86% of respondents scoring below 3 most did not believe that KM activities were inhibited by the various languages spoken in the construction project environment.

A Kruskal-Wallis test and comparison of means between this factor and the KM Maturity Level Indicator was completed (as shown in Table 22 and Table 24). The Kruskal-Wallis Test revealed a statistically significant difference in the Language factor across the KM Maturity Levels, x^2 (6, n = 79) = 16.88, p = .01 (Pallant, 2011). A negative trend was found (M=-0.17) when the mean score per KM Maturity Level was examined with the less mature organisations scoring higher. Interestingly it must be noted the high score for level 4 and the very low scores for levels 5 and 6. Based on a crosstabulation analysis between this barrier and the KM Maturity Level Indicator (as shown in Table 34 in Addendum C) the breakdown of the 24.1% of respondents who considered this to be a barrier was as follows:

- 35.5% of respondents from an organisation with KM Maturity Level 0.

- 13.3% of respondents from an organisation with KM Maturity Level 1.
- 12.5% of respondents from an organisation with KM Maturity Level 2.
- 20.0% of respondents from an organisation with KM Maturity Level 3.
- 44.4% of respondents from an organisation with KM Maturity Level 4.

Therefore, this factor shows no significant impact on any KM maturity level and will not likely inhibit KM activities internally and from extending across organisational borders in the more mature organisations.

Further analysis was conducted for this factor to determine if there was any correlation between the score developed for the factor and the respondents indicated 1st Language as shown in Table 19 below. The expectation could have been that respondents that did not list English as their 1st language would score this factor higher, based on a comparison of means analysis this trend does appear to be slightly evident with the exceptions of Afrikaans and Sesotho.

Table 19: Comparison of Means Between 1st Language & Language Factor

First Language	Language Factor Mean Score
Northern Sotho	4.0
Tsonga	3.5
Venda	3.3
Xhosa	3.3
Tswana	3.2
Zulu	3.0
Afrikaans	2.6
English	2.3
Sesotho	1.5
Not Specified	1.5

4.3.3.8 Analysis of the TMO Challenges Factor

The final and 8th ranked factor researched was the challenges faced by the modern Temporary Multi-Organisation coalitions formed for construction projects. Based on the feedback received from the respondents this factor resulted in a mean score of 2.4 with 68% scoring below 3. This score is well below the middle ground score of 3 and indicates that most respondents did not believe that the challenges faced in with a TMO impacted the KM activities.

A Kruskal-Wallis test and comparison of means between this factor and the KM Maturity Level Indicator was completed (as shown in Table 22 and Table 24). The Kruskal-Wallis Test revealed no statistically significant difference in the TMO Challenges factor across the KM Maturity Level Indicators, x^2 (6, n = 72) = 6.59, p = .36 (Pallant, 2011). Based on the percentage of respondents indicating this to be a barrier and the mean score per KM Maturity Level Indicator it was evident that overall respondents from most organisations did not view this as a barrier to KM, irrespective of their organisations KM maturity with only a slight negative trend found (M=-0.09). Based on a crosstabulation analysis between this factor and the KM Maturity Level Indicators (as shown in Table 31 in Addendum C) the breakdown of the 11.1% of respondents who considered this to be a barrier was as follows:

- 11.1% of respondents from an organisation with KM Maturity Level 0.
- 20.0% of respondents from an organisation with KM Maturity Level 1.
- 14.3% of respondents from an organisation with KM Maturity Level 2.
- 10.0% of respondents from an organisation with KM Maturity Level 6.

Therefore, this factor shows no significant impact on any KM maturity level and will not likely inhibit KM activities internally and from extending across organisational borders in the more mature organisations.

4.3.4 Opinion Towards KM and Extending KM Activities Beyond Organisational Borders

The analysis of the final objective of this research was split into first the analysis of the opinion towards KM and then to extending KM activities beyond organisational borders by the respondents who work in the South African construction sector.

4.3.4.1 Opinion Towards KM

With an overall mean score of 3.8 most respondents had a positive outlook towards Knowledge Management. A Kruskal-Wallis test and comparison of means between the Opinion Towards KM and the KM Maturity Level Indicator was completed (as shown in Table 22 and Table 24). The Kruskal-Wallis Test revealed a statistically significant difference in the scores for the Opinion Towards KM across the KM Maturity Level indicators, x^2 (6, n = 79) = 26.95, p = .001 (Pallant, 2011). Interestingly this positive outlook towards KM was not significantly skewed towards organisations with higher KM maturity when examining the mean score per KM Maturity level Indicator.

A further comparison of mean scores was completed to determine the outlook towards KM based on the respondents age, gender, ethnicity and region of operation. The results of this analysis showed that there was a uniform positive opinion amongst all age groups and gender. Interestingly when the mean score was compared per the participants ethnicity there was a noticeably lower opinion of KM by Black and Coloured respondents. Further organisations who operated in Gauteng, the Western Cape, the North West, Internationally and Nationally had an appreciably higher opinion on KM compared to the other regions as shown in Table 20 below.

Table 20: Additional Analysis for the Opinion Towards KM

Age of Respond-ent	Mean Score	Gender	Mean Score	Ethnicity	Mean Score	Region of Operation	Mean Score
18-24	4.0	Male	2.8	Asian	4.5	International	4.1
25-34	3.7	Female	3.9	White	4.1	Western Cape	4.0
35-44	3.7	Unknown	2.8	Indian	4.0	North West	4.0
45-54	3.8			Unknown	3.9	Unknown	4.0
55-64	4.1			Coloured	3.5	Gauteng	3.9
65+	3.8			Black	3.4	National	3.9
						KwaZulu Natal	3.4
						Eastern Cape	3.2
						Northern Cape	3.0
						Mpumalanga	3.0
						Limpopo	1.5

4.3.4.2   Opinion Towards Extending KM Activities Beyond Organisational Borders

With an overall mean score of 4.1 most respondents had a clear positive outlook towards extending KM beyond their organisation boundaries. A Kruskal-Wallis test and comparison of means between Objective 8 and the KM Maturity Level Indicator was completed (as shown in Table 22 and Table 24). The Kruskal-Wallis Test revealed no statistically significant difference in the scores for this opinion across the KM Maturity Levels, x^2 (6, n = 66) = 9.37, p = .15 (Pallant, 2011). When examining the mean score per KM Maturity Level there is only a slight positive trend evident (M=0.06) between the variables.

A further comparison of mean scores was completed to determine the outlook towards KM based on the respondents age, gender, ethnicity and region of operation. The results of this analysis showed a uniform positivity for all age groups, gender and ethnicity. The International, National and Kwa-Zulu Natal regions scored well above the overall average whilst the Northern Cape, North West and Gauteng regions scored below the average as shown in Table 21 below.

Table 21: Additional Analysis for the Opinion Towards Extending KM Activities Beyond Organisational Borders

Age of Respondent	Mean Score	Gender	Mean Score	Ethnicity	Mean Score	Region of Operation	Mean Score
18-24	4.0	Male	4.1	White	4.1	National	4.3
25-34	4.1	Female	4.0	Black	4.1	International	4.2
35-44	4.1	Unknown	3.7	Unknown	4.0	KwaZulu Natal	4.2
45-54	4.1			Asian	4.0	Eastern Cape	4.0
55-64	4.0			Coloured	3.5	Western Cape	4.0
65+	4.1					Gauteng	3.8
						North West	3.7
						Unknown	3.7
						Northern Cape	3.0

4.3.5 Additional Questions

The analysis of the additional questions asked in order to gain supplementary information was as follows.

4.3.5.1 Organisations Competitiveness and Extending KM beyond Organisational Borders

Based on the feedback shown in Table 17 for Question 64, 42% of the respondents indicated that they believed their organisations competitiveness would be diminished with a further 20% of respondents indicating a neutral response. Only the remaining 38% of respondents indicated that they didn't believe their organisations competitiveness would be diminished. A Kruskal-Wallis test and comparison of means between this question and the KM Maturity Level Indicator was completed (as shown in Table 22 and Table 24). The Kruskal-Wallis Test revealed no statistically significant difference in the scores for the question across the KM Maturity Levels, x^2 (6, n = 65) = 6.72, p = .35 (Pallant, 2011). The comparison of means showed a very slight positive trend (M=0.04) towards the less mature organisations believing their competition would be diminished whilst more mature organisations were less inclined to think this.

4.3.5.2 Continuity of the Project TMO Over Multiple Projects

Based on the feedback for Question 71 shown in Table 17 only 6% of respondents indicated that continuity in TMO's over successive projects would not improve project performance with 14% of respondents being neutral. Therefore, with an overall mean score of 4.02, the remaining 80% of respondents indicated that this would improve project performance. A Kruskal-Wallis Test (shown in Table 22) revealed a statistically significant difference in the response to the question across the KM Maturity Levels, x^2 (6, n = 66) = 15.26, p = .018 (Pallant, 2011). By examining the distribution of mean scores per KM Maturity Level (as shown in Table 24) a very slight positive trend was found (M=0.05).

Based on the feedback for Question 76 shown in Table 17 only 6% of respondents indicated that project performance had not been impacted by knowledge/experience leaving the country with 22% of respondents being neutral. The overall mean score of 4.0 was driven by the 72% of respondents that agreed that project performance had been impacted. A Kruskal-Wallis Test (as shown in Table 22) revealed no statistically significant difference in the response to this question across the KM Maturity Levels, x^2 (6, n = 65) = 5.05, p = .538 (Pallant, 2011). The examination of the mean score per KM Maturity level (as shown in Table 24) showed only a small positive trend (M=0.11).

4.3.5.4   Disciplinary Action

Based on the feedback for Question 77 shown in Table 17 26% of the respondents indicated that disciplinary action would likely take place if they shared knowledge outside their organisation beyond their contracted scope. 30% of respondents were neutral and the remaining 44% indicated that no disciplinary action would be taken. A Kruskal-Wallis test and comparison of means between this question and the KM Maturity Level Indicator was completed (as shown in Table 22 and Table 24). The Kruskal-Wallis Test revealed no statistically significant difference in the scores for the question across the KM Maturity Levels, x^2 (6, n = 66) = 7.15, p = .31 (Pallant, 2011). The comparison of means showed that there was only a slight positive linear relationship between the variables (M=0.07).

Table 22: Results of All Kruskal-Wallis Tests

Item 1	Item 2	Kruskal-Wallis H	df	Asymp. Sig.
Turnover of Organisation	KM Maturity Level Indicator	1.213	5	.944
Key Barrier 1	KM Maturity Level Indicator	31.018	6	.001
Key Barrier 2	KM Maturity Level Indicator	24.469	6	.001
Key Barrier 4	KM Maturity Level Indicator	16.011	6	.014
Key Barrier 5	KM Maturity Level Indicator	12.046	6	.061
KM Training Factor	KM Maturity Level Indicator	18.459	6	.005
Job Security Factor	KM Maturity Level Indicator	15.130	6	.019
TMO Challenges Factor	KM Maturity Level Indicator	6.590	6	.360
Continuity Factor	KM Maturity Level Indicator	9.733	6	.136
Management Support Factor	KM Maturity Level Indicator	15.039	6	.020
Language Factor	KM Maturity Level Indicator	16.878	6	.010
Contractual Framework Factor	KM Maturity Level Indicator	4.100	6	.663
Opinion Towards KM	KM Maturity Level Indicator	26.949	6	.000
Opinion Towards KM	Region of Operation	16.248	9	.062
Opinion Towards KM	Age of Respondent	2.729	5	.742
Opinion Towards KM	Gender of Respondent	.013	1	.908
Opinion Towards KM	Ethnicity of Respondent	7.580	4	.108
Opinion Towards Extending KM Beyond Organizational Borders	KM Maturity Level Indicator	9.371	6	.154
Compromised Competition	KM Maturity Level Indicator	6.720	6	.348
Disciplinary Action	KM Maturity Level Indicator	7.145	6	.308
Opinion Towards Extending KM Beyond Organizational Borders	Region of Operation	7.621	8	.471
Project Team Continuity	KM Maturity Level Indicator	15.259	6	.018
Knowledge Leaving SA	KM Maturity Level Indicator	5.046	6	.538

Table 23: Results of All Spearman Correlation Analysis Completed

Item 1	Item 2	Spearman's rho	N	Sig.
Key Barrier 1	Question 34	.446	79	.000
Key Barrier 2	Question 39	.238	79	.035
Key Barrier 3	Question 60	.103	72	.390

Table 24: Result of All Comparison of Mean Tests vs the KM Maturity Level Indicator

Item	KM Maturity Level Indicator							M*
	0	1	2	3	4	5	6	
Key Barrier 1	3.1	3.2	3.6	3.4	4.0	4.6	4.1	0.22
Key Barrier 2	3.2	3.1	3.7	3.8	3.5	4.7	4.3	0.22
Key Barrier 4	3.0	3.3	4.4	2.5	4.2	5.0	3.9	0.25
Key Barrier 5	2.7	2.6	3.4	2.1	2.7	3.7	3.1	0.10
Time Factor	3.3	3.5	2.7	4.1	3.3	2.8	3.0	-0.06
KM Training Factor	3.6	2.9	2.5	2.6	2.7	2.0	2.2	-0.21
Management Support Factor	3.1	2.8	3.5	2.7	3.2	2.0	2.5	-0.13
Continuity Factor	3.1	3.0	2.6	3.5	2.8	2.7	2.4	-0.09
Contractual Framework Factor	2.9	2.9	2.4	3.0	2.8	2.7	2.8	-0.01
Job Security Factor	3.0	3.0	2.8	2.4	2.7	1.5	2.0	-0.22
Language Factor	2.9	2.3	2.4	2.6	3.1	1.5	1.6	-0.17
TMO Challenges Factor	2.5	2.7	2.1	2.0	2.0	2.0	2.2	-0.09
Opinion Towards KM	3.4	4.2	3.9	3.1	4.1	3.5	4.6	0.09
Opinion Towards Extending KM Activities beyond Organizational Borders	4.0	3.6	4.2	4.4	4.5	3.7	4.4	0.06
Compromised Competition	2.8	2.8	3.3	3.7	3.1	2.0	3.8	0.04
Project Team Continuity	3.7	3.8	4.1	4.3	4.7	3.0	4.5	0.05
Knowledge Leaving SA	3.7	4.1	3.9	4.3	4.1	4.0	4.6	0.11
Disciplinary Action	3.1	3.2	4.0	4.3	2.9	5.0	3.3	0.07

***M shows the linear trendline coefficient developed for each set of mean scores from the trendlines in Figure 1.**

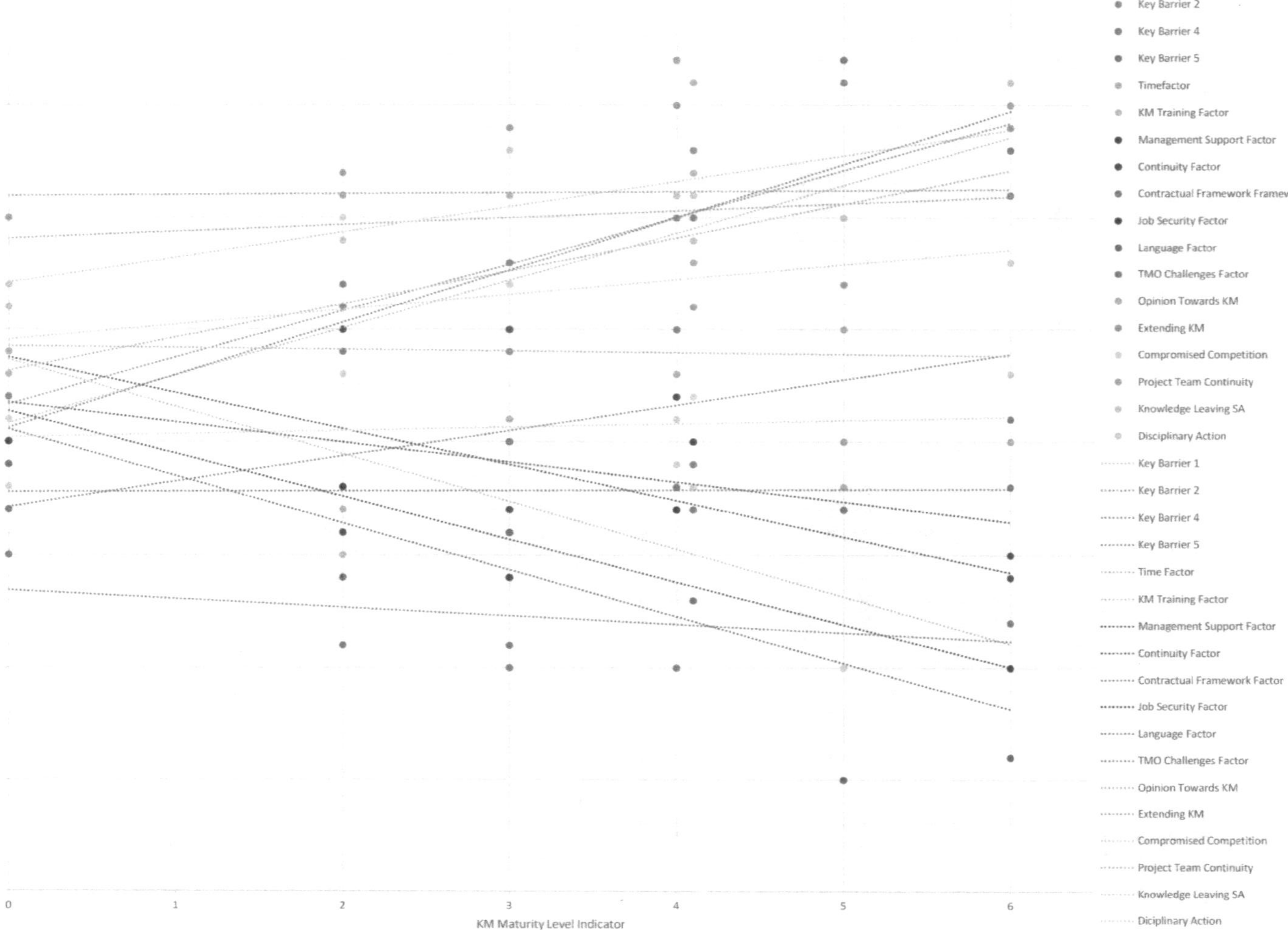

Figure 1: Scatter Plot of Results for the Comparison of Means Analysis with KM Maturity Level Indicator

4.4 Discussion of Findings

The research of Kruger and Johnson (2009) raised the question of why South African organisations do not extend their KM practices beyond their organisational borders without being able to provide a conclusion based in evidence. As discovered in the literature review of this research, it is when an organisation extends its KM activities beyond its own organisational borders that the value of KM can be maximised. According to Kruger and Snyman's (2007) KM maturity model this type of KM activity occurs when an organisation has reached the 6th level of KM maturity.

The following section of this research discusses the results of the data generated by the survey and the analysis completed in relation to the review of the literature previously discussed. This research had 4 objectives to address with the first being to determine a KM maturity level for each respondent's respective organisation, followed by an analysis of key barriers that were thought to be possible inhibitors to KM activities and extending KM activities beyond organisational borders in South Africa. The third objective of the research was to develop a ranked list of the factors behind these key barriers to determine the severity and prevalence of these factors in inhibiting KM activities and extending KM activities beyond organisational borders. The final objective was to determine the current opinion towards KM and extending KM activities beyond organisational borders of the South African construction sector.

4.4.1   KM Maturity Level Indicator

The initial step taken by this research was to determine the KM maturity level of each organisation that participated in the survey. This indicator was developed utilising feedback received from each respondent on select question which related back to each level of KM maturity developed by Kruger and Snyman (2007). The results developed showed that most organisations that responded had low KM maturity with a significant proportion (43%) not even qualifying for a KM maturity level of 1. The full breakdown of how many organisations qualified for each maturity level is shown in Table 7. As this research is focussed on the topic of extending KM activities beyond organisational borders it was encouraging to find that 10 organisations (11%) qualified for level 6 and therefore indicated that their KM activities were extending beyond organisational borders currently. As noted in the analysis the results of this research did not corroborate Kruger and Snyman (2009) findings that KM maturity and its growth was more prevalent in larger organisations. This result may be due to the small sample frame size of respondents to this research or there could have been growth in KM maturity in SME's in the 10 years since the publication of their results. Whilst this result must be noted it was beyond the scope of this research to examine this further.

4.4.2   Key Barriers

The second objective of this research was to determine the perception of the respondents towards the impact the identified key barriers had on inhibiting KM activities and extending these activities beyond organisational borders.

4.4.2.1   Key Barrier 1: KM Training

The first barrier to be examined by this research was to determine whether enough training was given to employees regarding KM. As discussed in the literature review KM activities can often be overwhelming for employees. Without the necessary training on KM and guidance on how to complete the activities this can be a barrier to a successful KM strategy (Schindler and Eppler, 2003; Shokri-Ghasabeh and Chileshe, 2014). The analysis concluded that most respondents believed they received enough training and a trend was found that this barrier was less prevalent in mature organisations. Therefore, it is less likely to be prevalent in inhibiting KM activities internally and from extending beyond organizational borders in mature organisations.

All 16.5% of respondents who indicated that they did not receive enough training were from organisations that had a low KM maturity level of 0 or 1. However, the mean scores for the less mature organisations were all still above neutrality and therefore indicated that, on average, they received enough training. This is an interesting development and could be explained by the thought that the respondents of the less mature organisations would have a limited understanding of KM and therefore did not know what they did not know. Alternatively, the respondents truly had received enough training in KM, but the maturity of their organisation has been stifled by other factors. An examination of these factors further was however beyond the scope of this research.

Further, as the literature review demonstrated, the construction sector is still predominantly informal globally and resources predominantly rely on their experience gained through on the job training (Carrillo and Chinowsky, 2005; Maqsood, Finegan and Walker, 2006). When asked if they believed that lessons could not be learned through training manuals and had to be learnt on the job 55% of respondents indicated they agreed with a further 19% being neutral (as shown in Table 17). This result is then a clear flag and indicates that the way KM training is being carried out is not effective.

Building on the topic of KM Training, as shown in Table 17, most respondents firmly believed that project performance in the local sector had been impacted by knowledge and experience leaving the South Africa.

4.4.2.2   Key Barrier 2: Organisational Culture

The culture of an organisation is critical to how effectively knowledge can be shared and the literature has shown how various cultural factors can be inhibitors. This research examined the most prominent factors that could impact an organisations culture in South Africa such as, fear of failure, conflict, job security uncertainty, internal competition and BBBEE type initiatives. Multiple researchers have explored how the culture of an organisation can impact a KM strategy and the culture has further been identified as an underlying factor to many other barriers that exist in the literature (Schindler and Eppler, 2003; Carrillo, Ruikar and Fuller, 2013; Shokri-Ghasabeh and Chileshe, 2014). Based on analysis completed by this research most respondents indicated that their organisations provided a conducive culture for sharing knowledge. The analysis revealed that the less mature organisations scored just above neutrality and the more mature organisations indicated clear agreement on the topic as shown in Table

24. The high score and trend found indicated that this barrier was less likely to be prevalent in inhibiting KM activities in most organisations and not likely to inhibit KM activities from extending across borders in the more mature organisations. Interestingly the breakdown of respondents who did not believe their environment was conducive to knowledge sharing was not limited to the less mature organisations as would be expected. It was found that and appreciable number of respondents (12.5% – 35.5%) from levels 0 through to 4 indicated that their environment was not conducive to sharing knowledge which is of concern. This result shows that there still organisations struggling to create the correct culture to share knowledge but as per the correlation analysis completed, as shown in Table 23, KM was having at least a small positive impact on this.

4.4.2.3   Key Barrier 3: Contractual Framework

The issue of whether the contractual framework of construction projects was inhibiting knowledge sharing was then examined. Research by Latham (1994) found that UK construction organisations had developed defensive relationships which had influenced the contractual frameworks entered into. Projects sponsors have since begun to address this issue by including continuous improvement strategies and collaborative clauses into contracts. Further, long-term relationship contracts, partnerships and learning agreements are being promoted, to enable feedback loop cycles to be developed both within internally and between organisations (Egbu, Sturges and Bates, 1999; Orange, Burke and Boam, 2000; Carrillo and Chinowsky, 2005; Carrillo, Ruikar and Fuller, 2013).

The result of this researches analysis indicated that most respondents did not believe the contractual framework was inhibiting knowledge sharing. However, when the breakdown of the results was examined it was found that only 15% of respondents clearly indicated that the contractual framework was conducive to sharing knowledge and 33% did not believe this was the case. The overall mean score was heavily influenced by most respondents indicating neutrality towards the topic. This result could therefore be influence by a large proportion of respondents not having enough knowledge on the contracts utilised in the construction industry and therefore not being able to provide positive or negative feedback to the questions asked. An examination of this further was however beyond the scope of this research.

To explore this issue further, when asked if the threat of penalties inhibited collaboration, roughly a third of respondents indicted agreement, a third were neutral and the remaining third disagreed. This result points to there still being room for improvement when establishing construction project contracts that foster and encourage sharing knowledge.

4.4.2.4   Key Barrier 4: Incentivisation

The topic of whether organisations were incentivising their employees to share knowledge in addition to their normal project activities was then examined. Researcher have found that resources lack the motivation to share knowledge (Kruger and Johnson, 2009c) and are sometimes resistant to learning from others (Schindler and Eppler, 2003). The motivation of employees to share knowledge is one of the many barriers that is clearly linked to the culture of the organisation. Further, the motivation and incentivisation of employees to share knowledge, will be linked to the maturity of the KM activities undertaken. Well-developed and clearly defined activities can be easily monitored, managed and therefore incentivised, whilst informal discussions are more difficult to evaluate and therefore incentivise.

The results generated by this research show that most respondents indicated that their organisations are incentivising their resources to share knowledge. The analysis confirmed that respondents from less mature organisations would more likely not agree with this positive feedback, but it was interesting to find that 10% of respondents from level 6 organisations provided negative feedback. This lends credibility to the thought that incentivising methods to employees is unique to each organisation and therefore even mature organisations could fall short in this regard due to other policies and/or factors. The overall score which indicates agreement towards being motivated to share knowledge and the positive trend indicate that this barrier was less likely to inhibited KM activities internally and from extending across organisational borders in the more mature organisations.

4.4.2.5   Key Barrier 5: Time

The time allocated to KM activities has long been recognised as a barrier to KM's effective implementation and is regularly found to be the most severe or prominent barrier (Schindler and Eppler, 2003; Shokri-Ghasabeh and Chileshe, 2014). The result of the analysis completed by this research show that most respondents agreed that not enough time was allocated. The negligent positive relationship between this barrier and KM maturity levels, with a low overall score developed indicates that the barrier is an issue regardless of KM maturity. This barrier will inhibit KM activities from extending across organisational borders even in the more mature organisations. This result coupled with the strong positive feedback specific to Question 57 (as shown in Table 17), where respondents were queried on if time pressure in projects prevents KM activities from taking place, lends credibility to the thought that KM activities are still not prioritised sufficiently.

4.4.3 Underlying Factors

The third objective of this research was to determine the prevalence and severity of factors that would have an influence on the key barriers identified.

4.4.3.1   Time Factor

Based on the feedback received (as shown in Table 14) it was found that Time was the highest ranked factor researched to be inhibiting KM's implementation. This would therefore certainly be a barrier which restricts KM growth and extending KM activities beyond organisational borders as this barrier was found to be significant across all KM maturity levels.

4.4.3.2   KM Training Factor

The second highest ranked factor was KM Training with close to half of the respondents indicating that this factor was an issue. A relationship was established showing that this barrier was more prevalent in the less mature organisations. However, as there were still 20% – 30% of respondents indicating that this factor was a barrier in the more mature organisations, training in all organisations should be prioritised. As this factor ranked high in the list developed and its evidence to be an issue in across the board the KM training barrier is certainly inhibiting KM maturity growth, especially in the less mature organisations. However, the factor was found to not be prevalent in the more mature organisations and would therefore be less likely to inhibit KM activities from extending across organisational borders.

4.4.3.3   Management Support, Continuity and Contractual Framework Factors

Following the KM Training factor in ranked order was "Management Support", "Continuity" and the "Contractual Framework" factors. These 3 factors scored similar to KM Training with close to half of the respondents indicating the factor to be an issue. This result ties back to the research discussed in the literature review where Schindler and Eppler (2003) found managements support of KM strategies to be critical to its success, whilst Orange, Burke and Boam (2000) had identified how a lack of continuity in project team and the contractual frameworks utilised can inhibit knowledge transfer. Both the Management Support and Continuity factors had a slight negative relationship to KM maturity and were therefore more prevalent in less mature organisations and would certainly be impacting KM's growth in these organisations.

The respondents to this survey were greatly in favour of continuity of project teams over multiple projects with 80% indicating the believed this would improve project performance. The Contractual Framework scores were uniform across the maturity levels which was to be expected as discussed previously where it was found that as contracts are similar for most construction projects and therefore the KM maturity level wouldn't impact the opinion towards how the contracts limited knowledge transfer.

4.4.3.4   Job Security, Language and TMO Challenges Factors

The final 3 factors discussed in ranked order were; "Job Security", "Language" and "TMO Challenges" who scored below 3 and were thus not indicated as barriers by most respondents. The result of this analysis is encouraging as the literature had indicated that the diverse and competitive environment, as found in South Africa, could foster the development of these barriers (Ofori, Hindlet and Hugo, 1996; Kruger and Johnson, 2009c; Carrillo, Ruikar and Fuller, 2013).

4.4.4 Opinion Towards KM and Extending KM activities Beyond Organisational Borders

Kruger and Johnson's (2009) research showed that there had been growth in KM in the South African construction sector. Based on the analysis completed by this research this positive opinion towards KM was still prevalent currently and was not limited to the more mature organisations as shown in Table 24. Based on the analysis even organisations that did not qualify for a maturity level of 1 still had an overall positive opinion towards KM. This result would indicate that the strength and value KM holds has been recognised but its implementation is being inhibited by multiple possible factors. Kruger and Johnson (2009) stated that as the growth and uptake in KM had been predominantly seen in larger organisations, it was the availability of funding and resources that would inhibit KM's implementation in South African organisations.

Further analysis completed showed that this positive opinion towards KM was not dependent on the respondents age or gender. However, noticeably lower mean scores were found for Black and Coloured respondents when the results were broken down per ethnicity. Further when the opinion of KM was broken down per region of operation it was found that organisations that had a presence Internationally, Nationally, in the Western Cape or Northern Cape scored significantly above average. However. it must be noted that these final two findings were found to not be statistically significant as shown in Table 22. Therefore, whilst these findings are not definitive, they do indicate an avenue for future research to investigate as it is interesting to note the possible link between a positive opinion towards KM and organisations that have an international link. This could be driven by global corporate policies or strategies that have filtered through to the South African divisions.

With the recent upswing in collaborative partnerships being formed in order to improve knowledge transfer over organisational borders (Bakker *et al.*, 2011; Carrillo, Ruikar and Fuller, 2013) this research examined what the current opinion was from its respondents towards extending KM activities beyond their organisational borders. The result of the analysis completed showed that most respondents had a clear positive opinion towards this topic across all KM maturity levels. Further analysis showed this result was not dependent on age, gender and ethnicity. However, when broken down per region of operation there was a dip in the results for Northern Cape, North West and Gauteng. This result was however not found to be statistically significant as shown in Table 22.

The positive response to extending KM activities beyond organisational borders would appear to contradict the feedback Kruger and Snyman (2007) received in their survey where the respondents indicated that there was little motivation and drive to share knowledge. Further they found that there was a perception that this would decrease their organisations competitiveness. The respondents to this researches survey were asked this then directly and based on the analysis completed in Table 17 there was a near equal agreement and disagreement that their organisations competitiveness would be diminished by extending KM activities beyond their borders.

This result therefore indicated that whilst there exists a strong positive opinion of KM and towards extending KM activities across organisational borders there is still a perception that the organisations competitiveness could be impacted by doing this. This result points towards the need for organisations to then better understand how to set up collaborative agreements and partnerships that will protect each organisations core IP, whilst enabling the sharing of knowledge which can help each other improve. Encouragingly it was found that the respondents indicated in Table 17 that predominantly organisations would not currently take disciplinary action for sharing knowledge across organisational boundaries.

4.4.5 Discussion of Additional Findings

When asked what prevalent construction issues could be addressed by organisations working closer together and transferring knowledge freely the respondents indicated that they believed that "Coordination of Information" and "Planning & Scheduling" were the most likely to be improved as shown in Table 18. These are both issues that Durdyev and Hosseini (2019) identified as common causes of delays on construction project globally.

5. Summary of Findings, Conclusions and Recommendations

5.1 Introduction

To aid the South African construction sector to improve upon its collective KM maturity, there was a need to examine the barriers faced by construction organisations when trying to extend KM activities beyond organisational borders. The literature has shown that by highlighting the barriers that inhibit KM's implementation, organisations can overcome the challenges faced.

This research adopted a quantitative approach that reviewed the existing literature on the topic of barriers inhibiting Knowledge Management's implementation in the construction sector. Based on the knowledge gained through the literature review a survey instrument was developed and sent out to construction professionals in the South African construction sector. This q survey instrument tested the respondents on their perceptions of barriers to KM implementation in the South African construction sector.

5.2 Summary of the Findings

The following section summarises the findings made in the research in respect to each objective set out for analysis by this research report.

5.2.1 KM Maturity Level Indicator

The research completed found that 43% of the organisations that were represented in the survey did not qualify for a KM maturity level of 1 with the next 31% of respondent's organisations scoring 1 or 2 for KM maturity. Therefore 74% of the respondent's organisations that participated in the survey indicated that they had no, or an immature, KM strategy in place. This result is of concern for the validity of the results generated by this research as a broader range of respondents would provide more comprehensive results. However, it was encouraging to find that 11% of respondent's organisations indicated a KM maturity of level 6 and are therefore potentially maximising their knowledge management strategies and extending KM activities beyond organisational borders.

5.2.2 Key Barriers

This research then examined five key barriers that were identified in the literature to be potential inhibitors to KM activities and extending these activities beyond organisational borders in South African construction organisations. It was found that the key barrier of "allocation of time for KM activities" was a clear inhibitor to completing KM activities and as this barrier was identified to be an issue across all levels of KM maturity, it would be an inhibitor to extending KM activities beyond organisational borders.

It was found that the key barriers of "incentivising employees to share knowledge", "KM training", "organisation culture" and "contractual framework" were not clearly indicated as inhibitors overall by the respondents to completing KM activities or to extending KM activities beyond organisational borders.

The result of this analysis was however not in contradiction to the literature from where these barriers were found. The resultant analysis for these barriers showed an overall neutrality towards the barriers as opposed to clear agreement or disagreement.

5.2.3 Underlying Factors

This research found that when the key barriers identified for analysis were broken down into specific underlying factors that the issue of "Time" was found to be the highest ranked factor and again the only factor clearly identified as an inhibitor to KM activities and extending these beyond organisational borders.

The factors of "KM training", "Management Support", "Continuity" and "Contractual Framework" scored close to neutrality and were therefore not clearly identified inhibitors to KM activities and extending these beyond organisational borders. The breakdown of results showed that "KM Training", "Management Support" and "Continuity" were clearly inhibitors in the less mature organisations. However, in the more mature organisations, the scores for these factors were not clearly indicated as inhibitors, but there was enough negative feedback to indicate a concern across all KM maturity levels.

The statistically significant differences in scores for "KM Training" and "Management Support" per KM maturity level and clear improvement in scores for the mature organisations indicates that these inhibitors not only inhibit KM activities but also KM's maturity growth in the less mature organisations. Whilst the "Continuity" factor was not clearly defined as an inhibitor to KM activities, feedback showed that the respondents believed that project performance overall could be improved with greater continuity in the make-up of project teams.

The "Contractual Framework" factor was scored uniformly across all KM maturity levels positively and a breakdown of the results showed no clear indication that this factor inhibited KM activities or its growth.

The final three factors of "Job Security", "Language" and "TMO Challenges" were ranked lowest in the analysis completed and through an examination of their scores were found to be the least likely inhibitors to KM activities. South Africa is a diverse and challenging environment that struggles with many problems and so to find that the respondents did not identify the need to limit their knowledge transfer in order to protect their jobs is heartening. Further as South Africa has 11 official languages it is encouraging to find that this does not inhibit knowledge transfer in the project environment. The final factor examined and found to not be an inhibitor to KM activities and its growth related to the challenges faced in assembling project teams together to transfer knowledge. This is an interesting result as the modern construction projects are being completed by more geographically dispersed project teams. The South African construction sector is therefore clearly finding solutions to overcome this challenge.

5.2.4 Opinion Towards KM and Extending KM Activities Beyond Organisational Borders

It was found that overall the respondents had a positive opinion towards KM which was not limited to the more mature organisations. It was found that even organisations that had not qualified for the initial

KM maturity levels still had a positive outlook towards KM, which would indicate that the value KM can deliver has been recognised in South Africa construction organisations, but its implementation was being inhibited. This research was able to show that the positive opinion towards KM was not dependant on age or gender, whilst there was a dip in the still positive opinion of Black and Coloured respondents. An interesting observation which should be explored in further research was that the opinion of KM per region of operation was found to be higher in organisations that operated nationally and internationally.

Building on the positive opinion towards KM found this research established that there was a clear positive opinion towards extending KM activities beyond organisational despite the analysis completed showing near equal agreement and disagreement to the perception that the competitiveness of the organisation would be negatively impacted.

5.3 Conclusions

Based on these findings, this research concludes that a clear positive opinion of KM and extending KM activities beyond organisational borders in the South African construction sector exists. It was found that the insufficient allocation of time to KM activities was a significant inhibitor to KM's implementation and extending these activities beyond organisational borders. The amount of training given on KM and the support of management was found to be of concern in organisations with low KM maturities and it was established that the respondents believed that project performance would be improved if there was continuity in the make-up of project teams over multiple projects.

Whilst the contract framework utilised in construction projects was not clearly identified as a barrier it was found that there was a concern towards the competitiveness of an organisation being impacted by sharing information beyond organisational borders. This perception is therefore a discovered additional barrier to extending KM activities beyond organisational orders alongside the barrier of "Time" that needs to be overcome.

There was positive feedback to overcoming the TMO Challenges that are becoming prevalent in the modern project environment and the number of languages spoken in South Africa does not appear to be inhibiting the sharing of knowledge. Encouragingly it was found that despite an indication that construction project performances had been impacted by skilled professionals emigrating from South Africa there was no clear indication by the respondents that they were limiting their knowledge transfer to project their own jobs.

5.4 Recommendations of the Study

The findings show that KM practitioners need to ensure that KM activities are prioritised and that enough time is allocated for the activities to be efficiently completed. Further, organisations need to be cognisant that training is a continuous process and that irrespective of KM maturity training should be an ongoing process for all employees. The lack of managements support could potentially inhibit a KM strategy and where possible continuity in the make-up of the project team should be facilitated both within an organisation and in TMO's. This can be facilitated through the development of partnerships and long-term contractual relationships which seek to promote knowledge sharing whilst still ensuring that each organisations core IP is protected. By establishing these relationships, the perception that competition will be negatively impacted can be overcome and KM activities can extend beyond organisational borders freely. The evidence found that South African construction professionals are working together and share knowledge despite the adversity, diversity and other complex challenges found in the local environment is very encouraging and should be applauded.

Through the completion of this research a few areas where further research is needed were identified. Whilst it was beyond the scope of this research there would be value to completing a thorough examination of the strategies implemented by South African construction organisations to accurately plot their KM maturity against Kruger and Snyman's (2007) model and then determine if a relationship still exists today between the KM maturity level and the organisations size as found by Kruger and Johnson (2009a) in their research a decade ago.

The next topic that should be researched further was the interesting result found where respondents from organisations that were found to have a low KM maturity indicated that they received enough training on the topic of KM. This result could be argued to be due to respondents from the low maturity organisations not knowing what they do not know. Further research into this could highlight how SME's in South Africa struggle to provide adequate training to their employees on improvement strategies such as KM.

A further topic that should be researched was the result found that respondents did not believe that the contractual framework limited the knowledge transfer in construction projects in South Africa. As this topic is the subject of research globally and has been found to require attention by research by Orange, Burke and Boam (2000) the indication that this is not an issue in South Africa requires further research.

The final topic found to require further research was the result that the opinion towards KM was more positive for organisations that had an international or national presence over those who were limited to operate in only one province of South Africa. Research into whether the influence of global corporate strategies and policies is driving change in South African construction organisations could yield interesting results and should be explored further.

5.5 Contribution of the Study to Knowledge

This research contributes to the current body of knowledge on KM and its growth in South Africa. The literature review identified multiple factors that could have be inhibiting factors to the KM activities undertaken by construction organisations and further limit these activities from extending beyond organisational borders. Through the distribution of the survey instrument this research potentially provided knowledge on KM to the respondents reached and furthered their understanding of the value KM can unlock. Through the engagement and results generated this research has the potential to drive change within the construction sector by outlining the factors that are inhibiting KM from extending beyond organisation borders.

This research was limited to the feedback received from construction professionals in the South African construction sector and provides insight exclusively to this environment. However, the result found that time was a significant inhibitor to KM activities ties into research conducted globally. Further the results determined for other barriers such as "Management Support", "Continuity", "Contractual Framework", "Job Security", "Language" and "TMO Challenges" was not complete dismissal of these factors as barriers but rather that they were not clear inhibitors in the South African construction sector.

5.6 Limitations of the Study

This research acknowledges that it was limited first, and foremost, by the its scope of being conducted in South Africa and exclusively open to only construction professionals who had an operational presence in South Africa. The response rate for the research was low and with only 93 valid data sets in the sample frame for statistical analysis the results for this research cannot claim to represent whole construction sector of South Africa. Further most respondents to the survey were from small and micro organisations who had low KM maturities.

Due to time constraints the survey was only open for responses for 5 months and without the buy-in from most construction bodies in South African construction sector to provide the contact details of their members the survey did not reach as many professionals as was hoped for.

On review it is acknowledged that certain questions in the survey were not concise enough and enabled the low reliabilities found in the statistical analysis when developing the scores per barrier. Knowledge Management is a broad and all-encompassing topic when discussing organisational process and procedures. The development of a Pilot Study could have identified the flaws in the survey design and therefore ensured that the results of the survey were more concise and reliable.

This study should therefore be view as an indicator for where further and more extensive research should be conducted on the topic of Knowledge Management in the South African construction sector. The professionals who complete construction projects in this environment face some of the most challenging conditions and the insight of how they overcome these should be shared to KM practitioners globally.

References

A Guide to the Project Management Body of Knowledge (PMBOK Guide). 4th Editio (2008). Newtown Square, PA: Project Management Institute.

Bakker, R. M. *et al.* (2011) 'Managing the project learning paradox: A set-theoretic approach toward project knowledge transfer', *International Journal of Project Management*, 29(5), pp. 494–503. doi: 10.1016/j.ijproman.2010.06.002.

Baloyi, L. and Bekker, M. (2011) 'Causes of construction cost and time overruns : The 2010 FIFA World Cup stadia in South Africa', *Acta Structilia*, 18(1), pp. 51–67.

Bhargav, D. and Lauri, K. (2009) 'Collaborative knowledge management—A construction case study', *Automation in Construction*, 18(7), pp. 894–902. doi: 10.1016/j.autcon.2009.03.015.

Carrillo, P. (2005) 'Lessons learned practices in the engineering, procurement and construction sector', *Engineering, Construction and Architectural Management*, 12(3), pp. 236–250. doi: 10.1108/09699980510600107.

Carrillo, P. and Chinowsky, P. (2005) 'Exploiting Knowledge Management: The Engineering and Construction Perspective', *Journal of Management in Engineering*, 22(1), pp. 2–10. doi: 10.1061/(asce)0742-597x(2006)22:1(2).

Carrillo, P., Ruikar, K. and Fuller, P. (2013) 'When will we learn? Improving lessons learned practice in construction', *International Journal of Project Management*. Elsevier Ltd. APM and IPMA, 31(4), pp. 567–578. doi: 10.1016/j.ijproman.2012.10.005.

CIDB (2012) *The Construction Industry as a vehicle for Contractor Development and Transformation*. Pretoria.

CIDB (2019) *CIDB Construction Monitor - Employment; October 2019*.

CII (2015) *Performance Assessment 2015 Edition, Construction Industry Institute*.

Cooper, K. G., Lyneis, J. M. and Bryant, B. J. (2002) 'Learning to learn, from past to future', *International Journal of Project Management*, 20(3), pp. 213–219. doi: 10.1016/S0263-7863(01)00071-0.

Creswell, J. W. (2014) *Research Design: qualitative, quantitative and mixed methods approaches*. 4th edn. Thousand Oaks, California: SAGE Publications.

Durdyev, S. and Hosseini, M. R. (2019) 'Causes of delays on construction projects: a comprehensive list', *International Journal of Managing Projects in Business*. doi: 10.1108/IJMPB-09-2018-0178.

Egan, J. (1998) *The Egan Report - Rethinking Construction*. London.

Egbu, C., Sturges, J. and Bates, M. (1999) 'Learning from knowledge management and trans-organizational innovations in diverse project management environments.', *Association of Researchers in Construction Management*, 1(September), pp. 95–103. Available at: http://www.arcom.ac.uk/-docs/proceedings/ar1999-095-103_Egbu_Sturges_and_Bates.pdf.

Fellows, R. and Liu, A. M. M. (2012) 'Managing organizational interfaces in engineering construction projects: Addressing fragmentation and boundary issues across multiple interfaces', *Construction Management and Economics*, 30(8), pp. 653–671. doi: 10.1080/01446193.2012.668199.

Franco, L. A., Cushman, M. and Rosenhead, J. (2004) 'Project review and learning in the construction industry: Embedding a problem structuring method within a partnership context', *European Journal of Operational Research*, 152(3), pp. 586–601. doi: 10.1016/S0377-2217(03)00059-6.

Hardy, C., Phillips, N. and Lawrence, T. (2003) 'Resources, Knowledge and Influence: The Organizational Effects of Interorganizational Collaboration', *Journal of management studies*, 40(2), pp. 321–347.

Jacky, S. *et al.* (1999) 'Knowledge management and innovation: networks and networking', *Journal of Knowledge Management*, 3(4), pp. 262–275. doi: 10.1108/13673279910304014.

Kamara, J. M. *et al.* (2002) 'Knowledge management in the architecture, engineering and construction industry', *Construction Innovation*, 2(1), pp. 53–67. doi: http://dx.doi.org/10.1108/MRR-09-2015-0216.

Kruger, C. and Johnson, R. (2009a) 'Assessment of knowledge management growth: A South Africa perspective', *Aslib Proceedings: New Information Perspectives*, 61(6), pp. 542–564. doi: 10.1108/00012530911005517.

Kruger, C. and Johnson, R. (2009b) 'Determining the Value of Knowledge Management : A South Africa Perspective', *GlobDev*, (3), pp. 1–32.

Kruger, C. and Johnson, R. (2009c) 'Enablers of South African knowledge management maturity: Issues, principles and policies', *AMCIS 2009 Proceedings*, Paper 376. Available at: http://www.scopus.com/inward/record.url?eid=2-s2.0-84870371771&partnerID=tZOtx3y1.

Kruger, C. and Snyman, M. (2007) 'Guidelines for assessing the knowledge management maturity of organizations', *SA Journal of Information Management*, 9(3). doi: 10.4102/sajim.v9i3.34.

Latham, S. M. (1994) *Constructing the team*.

Maqsood, T., Finegan, A. and Walker, D. (2006) 'Applying project histories and project learning through knowledge management in an Australian construction company', *Learning Organization*, 13(1), pp. 80–95. doi: 10.1108/09696470610639149.

McCutcheon, R. T. (2003) 'Employment creation in public works', *Habitat International*, 19(3), pp. 331–355. doi: 10.1016/0197-3975(95)00001-v.

National Treasury: Republic of South Africa (2019) *Budget Review, National Treasury.* Available at:

http://www.treasury.gov.za/documents/National Budget/2019/review/Chapter 3.pdf%0Ahttp://www.treasury.gov.za/documents/national budget/2018/review/FullBR.pdf.

Ofori, G., Hindlet, R. and Hugo, F. (1996) 'Improving the Construction Industry of South Africa : A Strategy', *Habitat International*, 20(2), pp. 203–220.

Orange, G., Burke, A. and Boam, J. (2000) 'The Facilitation of Cross Organisational Learning and Knowledge Management to Foster Partnering within the UK Construction Industry'.

Pallant, J. (2011) *SPSS Survival Manual*. 4th edn. Allen & Unwin.

Peihua, Z. and Fung, F. N. (2013) 'Explaining Knowledge-Sharing Intention in Construction Teams in Hong Kong', *Journal of Construction Engineering and Management*, 139, pp. 280–293. doi: 10.1061/(ASCE)CO.1943-7862.

Rezgui, Y., Hopfe, C. J. and Vorakulpipat, C. (2010) 'Generations of knowledge management in the architecture, engineering and construction industry: An evolutionary perspective', *Advanced Engineering Informatics*. Elsevier Ltd, 24(2), pp. 219–228. doi: 10.1016/j.aei.2009.12.001.

Rivera, A. *et al.* (2017) 'Identifying the Global Performance of the Construction Industry', *53rd ASC Annual International Conference Proceedings*, pp. 567–575.

Schindler, M. and Eppler, M. J. (2003) 'Harvesting project knowledge: A review of project learning methods and success factors', *International Journal of Project Management*, 21(3), pp. 219–228. doi: 10.1016/S0263-7863(02)00096-0.

Senaratne, S. and Malewana, C. (2011) 'Linking individual, team and organizational learning in construction project team settings', *Architectural Engineering and Design Management*, 7(1), pp. 50–63. doi: 10.3763/aedm.2010.0133.

Senaratne, S. and Sexton, M. (2008) 'Managing construction project change: A knowledge management perspective', *Construction Management and Economics*, 26(12), pp. 1303–1311. doi: 10.1080/01446190802621044.

Senaratne, S. and Sexton, M. G. (2009) 'Role of knowledge in managing construction project change', *Engineering, Construction and Architectural Management*, 16(2), pp. 186–200. doi: 10.1108/09699980910938055.

Shokri-Ghasabeh, M. and Chileshe, N. (2014) 'Knowledge management: Barriers to capturing lessons learned from Australian construction contractors perspective', *Computing Handbook, Third Edition: Information Systems and Information Technology*, 14(1), pp. 108–134. doi: 10.1201/b16768.

Skyrme, D. and Amidon, D. (1997) 'Journal of Knowledge Management', *Journal of Knowledge Management*, 1(1), pp. 27–37. doi: https://doi.org/10.1108/JKM-07-2016-0274.

Terzieva, M. (2014) 'Project Knowledge Management: How Organizations Learn from Experience',

Procedia Technology. Elsevier B.V., 16, pp. 1086–1095. doi: 10.1016/j.protcy.2014.10.123.

Thwala, W. D. and Phaladi, M. J. (2009) 'An exploratory study of problems facing small contractors in the North West province of South Africa', *African Journal of Business Management*, 3(10), pp. 533–539. doi: 10.5897/AJBM09.122.

de Wet, P. (2019) *The definitions of micro, small, and medium businesses have just been radically overhauled – here's how*, Business Insider SA. Available at: https://www.businessinsider.co.za/micro-small-and-medium-business-definition-update-by-sector-2019-3 (Accessed: 24 November 2019).

Wiig, K. M. (1997) 'Knowledge Management: An Introduction and Perspective: Journal of Knowledge Management: Vol 1, No 1', *Jornal of Knowledge Management*, 1(1), pp. 6–14. Available at: http://www.emeraldinsight.com/doi/pdfplus/10.1108/13673279710800682.

Zou, P. X. W. and Lim, C.-A. (2002) 'The implementation of knowledge management and organisational learning in construction company', *Advances in Building Technology*. Elsevier, pp. 1745–1753. doi: 10.1016/B978-008044100-9/50215-1.